Feeling Like a Physicist

Chavan's UED research, Volume 10

Sandeep Chavan

Published by SANDEEP CHAVAN, 2025.

FEELING LIKE A PHYSICIST

First edition. May 27, 2025.

ISBN: 979-8231827268

Written by Sandeep Chavan.

Table of Contents

Prologue 1
1: The First Question 4
2: The Physics of Discomfort 9
3: The Syllabus Stops Here 14
4: Teacher vs Thinker 18
5: What If Nothing Moves? 22
6: Sitting With Silence 27
7: Ripples Over Rules 31
8: Birth of a Framework 36
9: When Force Felt Fake 40
10: Seeing Time Differently 45
11: Arguing With Einstein 51
12: Myth as Memory 55
13: Karma, But Not Religious 59
14: Alignment is Everything 63
15: Soul, Observer, Intelligence 69
16: Ignored, But Not Silent 76
17: No Lab, No Label, No Limits 80
18: My Family Didn't Read It, But Felt It 84
19: The 1000-Hour Window 88
20: Feeling Like a Physicist 92
21: Truth Has No Awards 97
22: The LOC Will Outlive Me 102
23: To the Quiet Thinkers 106
24: I Don't Want Followers 110
25: This Was Never About Physics 114
Closing Note 118
Epilogue: The Ripple Never Ends 120
Appendices 122

Appendix A: Physics 123
Appendix B: Cosmology 129
Appendix C: Philosophy 135
Appendix D: Psychology & Human Behavior 142
Further Reading 149
Acknowledgements 153
About the Author 155
Disclaimer 157

To my students — past, present, and future, You may not have read this book,but you were the field in which it formed.

Your questions, silences, frustrations, and insightsshaped the ripples that eventually became these words.

This isn't just dedicated to you —it was written because of you.

May you one day feel the physics behind your thoughts,the alignment behind your actions,and the truth behind every consequence.

Keep observing.Keep questioning.And above all... stay aligned.

— *Sandeep*

"Truth doesn't arrive through noise.It appears when alignment becomes impossible to ignore.Physics was never about equations —it was always about listening deeply enough to feel what never needed to be explained."

— Sandeep Chavan

Prologue

N*ot a Physicist. Just a Human Who Asked Too Many Questions.*

I never studied in a top university. Never worked in a lab with million-dollar instruments. Never wore a white coat or carried a research badge. I wasn't invited to a conference, cited in journals, or even welcomed by the so-called "community of knowledge."

But I kept asking questions.

The kind of questions that most people either ignore, fear, or laugh at. The kind that unravel your sleep, confuse your friends, and isolate you from every comfort zone you once trusted. And slowly, without permission or recognition, I started feeling like... a physicist.

Not because I knew the equations better. But because I began to see the universe differently. Not as a collection of particles and forces. But as a web of consequences, alignments, and intelligent ripple interactions.

This book isn't a theory. It's not a paper. It's not even a summary of all I've discovered — because truth can't be summarized. It can only be shared.

This is a *dialogue*. Between you and me. Between who I was when I began this journey — and who you might be as you begin yours.

In these pages, I'll share what it felt like to unlearn the known. To break open mainstream science and realize it wasn't wrong — just unfinished. To sit alone with a blank notebook and feel the echo of the cosmos whispering something far older than math.

I've spent over a thousand hours thinking, reflecting, writing, observing. I've redefined gravity, questioned time, dissected mythology, and explored the soul — not as a believer, but as a builder of thought. And every page I wrote was not to teach others — but to understand myself.

If you've ever stared at the sky and felt that equations don't do it justice... If you've ever sensed that your mind was designed not just to follow — but to *discover*... If you've ever doubted what you were taught because your own senses told a different story...

Then this book is for you.

I'm not here to convince you. I'm here to show you what it felt like — to be free from needing to convince anyone.

Because the moment you stop chasing validation... the universe starts speaking in your language.

Welcome to *Feeling Like a Physicist.* Let's begin.

Part I: The Quiet Unlearning

1: The First Question

"*What if they're all wrong — not out of malice, but out of method?*"

It didn't start with a eureka moment. No falling apple. No divine inspiration. Just a strange, creeping discomfort that began whispering in my thoughts like an unwanted guest. Quiet, persistent, and far too polite to be ignored.

I was teaching — again. The same concepts. The same laws. The same chalk-and-board physics that I had repeated year after year. Newton's laws. Equations of motion. Electromagnetic theory. Thermodynamics. Relativity. All of it. It wasn't that I doubted the content. I doubted the completeness of it.

Every time I said, "Force equals mass times acceleration," a part of me wanted to interrupt myself and say, "But is that really how nature thinks?" Does a tree grow because of a force? Does the Earth spin because an equation tells it to? Does light really travel — or are we just perceiving it as movement because we haven't looked deeply enough?

That was when the first question emerged: "What if they're all wrong — not out of malice, but out of method?"

What if these brilliant scientists, these mathematical magicians of past centuries, were not lying to us... They were simply limited. Limited by their instruments, their perceptions, their cultural beliefs, and — above all — by the language they used to describe what they saw.

They saw the world in parts, so they created parts. They saw interaction, so they called it force. They saw change, so they called it time.

And I… I was just repeating all of it. Mechanically. Faithfully. Blindly.

The discomfort grew slowly. It wasn't dramatic. It wasn't even visible to others. But it changed everything.

It made me pause between lectures. It made me linger longer at the whiteboard. It made me stare at simple diagrams for hours as if they were lying to me. And in some ways, they were.

Because they showed me what was convenient — not what was true.

I didn't have a lab. I didn't need one.

My lab became my own mind, and my equipment was my curiosity. That's when everything around me turned into an experiment. The stillness of a fan. The behavior of water. The way shadows changed during the day. The difference in people's emotions before and after a decision. The very act of "feeling" something.

I realized I wasn't just asking about physics. I was asking about reality.

The more I thought, the more I began to notice: Every law we believe in begins with an assumption. Every formula hides a compromise. Every theory is shaped not only by evidence — but by intent.

And that's when the second wave of questioning hit me:

What if nothing moves? What if movement itself is an illusion born out of ripple realignment?

Think about it. We don't actually *see* objects move. We see change in position over time, both of which are perceived values. Our perception aligns with certain changes, and we call that "motion." But what if, at the core, it's not motion — it's realignment? What if everything is just holding its place in a complex field of ripples, and any shift is not a journey, but a response to field imbalance?

It felt wild to think like this. But it also felt freeing.

You see, I wasn't trying to disprove physics. I was trying to understand why it felt incomplete.

Every system I had studied — from mechanical to quantum — was built on the assumption that we can model reality using measurable, force-driven equations.

But what if force itself is just an after-effect? A name we gave to the interaction we didn't fully understand?

I started seeing the universe not as a machine — but as a dynamic equilibrium. Where energy doesn't just flow — it seeks balance. Where matter isn't static — it's compressed ripple. Where time isn't a dimension — it's a perception delay due to misalignment.

Now, this wasn't easy. Let me be honest.

Thinking like this doesn't get you praise. It gets you confused stares, mocking smiles, and subtle reminders to "stick to the syllabus." Students want answers. Colleagues want consistency. Society wants certainty.

But truth... Truth wants only one thing from you: To stop lying to yourself.

And once I stopped lying, I couldn't unsee it.

I went back to my own notes, to the chapters I had once read as a student. Each time, I found something unsettling. Equations that solved problems — but didn't explain causes. Concepts that described effects — but skipped intent.

Take gravity, for instance.

We were told objects fall due to gravitational force. Fine. But why does this "force" exist? Einstein brought elegance — curvature of spacetime — but even that started with assumptions about space and time as dimensions. Why should space bend? Why does time stretch? What causes the curvature? And most of all — What is actually acting?

Is there really an "action" — or are we just perceiving energy alignment between two fields?

And here's what people don't tell you: You don't need to be a professor to think like this. You just need courage. Courage to not be validated. Courage to not be published. Courage to be wrong in public — and still right in your heart.

Because the universe doesn't care about your résumé. It cares about your alignment with truth.

And when I started aligning... Ideas didn't come — they arrived. Not through logic alone, but through a deep resonance. I could feel when something made sense beyond words. I could feel when a concept wasn't clicking because it was never meant to — it was just a workaround.

That's how I knew I wasn't thinking like a teacher anymore. I was feeling like a physicist.

That first question — "What if they're all wrong, not out of malice, but out of method?" — was not an accusation. It was an awakening.

It showed me that truth doesn't need an apology, but it does need attention. It needs someone to sit with it, quietly, sincerely, until it speaks back.

That's what I did. I sat with questions the world had stopped asking.

And somewhere in that silence, I found a voice that wasn't mine... But felt like home.

If you've ever felt uncomfortable with answers that everyone else accepts — stay with that feeling. It's not ignorance. It's the door.

And I promise you — behind that door, is a universe waiting to reveal itself, not through theory, but through truth that you were born to feel.

2: The Physics of Discomfort

Why doubt is often more real than the data How I began noticing inconsistencies in my own explanations

There is a strange thing about teaching: you can speak for hours without truly listening to yourself.

You walk into the classroom with a topic, a whiteboard, a set of diagrams, and a perfectly choreographed sequence of explanations. The students nod. Some take notes. Others drift. The syllabus advances. Everything moves.

Except one thing.

Your own certainty.

The first time I felt real discomfort while teaching physics, it wasn't because I didn't understand the topic. It was the opposite — I understood it too well.

I was teaching the second law of thermodynamics — entropy, the so-called arrow of time. I remember saying, "In any closed system, entropy tends to increase." And then I paused. Because something inside me whispered, "What system is ever truly closed?"

We draw imaginary boxes around systems to make calculations easy. We call them isolated, insulated, perfectly idealized. But in reality — there are no perfect boxes. Energy leaks. Influence bleeds. Nothing is untouched. So if our foundational law starts with an illusion — where does that leave the truth?

I shook off the thought and continued the class. But the discomfort stayed.

This was the beginning of a strange phase in my life. I started noticing more than I was explaining. I became hyper-aware of my own teaching voice — not what it was saying, but what it was skipping.

I was no longer just a teacher. I was an observer of my own beliefs.

Every time I explained gravity, I noticed how I rushed through the "why" to get to the "how." Every time I described time dilation, I cringed at how casually I said, "Time slows down," as if time were a thing that could speed up or down like a scooter.

What exactly is "slowing down"? Relative to *what*? What is time, really, if not just our perception of change?

I wasn't doubting science. I was doubting the language we used to describe it. The metaphors, the shortcuts, the assumptions dressed up as definitions.

Let me put it this way: If physics is the operating system of the universe, then doubt is the debugging tool.

And I was full of bugs.

I couldn't shake off the inconsistencies. Not because they were obvious — but because they were hiding in plain sight.

- We say light is a wave and a particle. But why is it allowed to be both when everything else must choose?
- We say energy is conserved — but in what field is it conserved, if the universe is expanding?

- We say matter curves space — but what curves matter?

These questions weren't in textbooks. They were in my gut.

And it's funny — you can ignore logic for years, but you can't ignore that tight knot in your chest when something doesn't sit right.

That knot was my real curriculum.

Some days, I felt like a fool. What right did I have to question Newton or Einstein or Feynman? I had no PhD. No lab. No affiliation. Just a marker, a classroom, and an uncomfortable feeling.

But I also knew something deeper: No scientist ever discovered truth by feeling comfortable.

The ones who changed the world didn't start with answers. They started with unease.

So I embraced my discomfort.

Instead of hiding it behind smoother lectures or well-rehearsed slides, I began leaning into it. I let it disturb me. I let it follow me home. I let it sit with me in silence, on nights when the world had gone to sleep, and I was still staring at my ceiling, wondering:

"What if we've been building elegant theories over cracked foundations?"

Because here's what no one tells you: Doubt isn't the opposite of knowledge. Doubt is the sign that knowledge is alive.

I began rewriting old concepts in my mind — not to reject them, but to rescue them.

I asked:

- What if gravity isn't a force, but a field tension trying to reach equilibrium?
- What if time doesn't flow, but emerges from perception delays during ripple misalignment?
- What if nothing truly moves — and what we call motion is just shifting energy between nested ripple fields?

These weren't answers. They were new discomforts. But this time, they didn't scare me. They felt honest.

And then came a strange turning point.

I realized that every time I felt discomfort while teaching, it wasn't a sign of failure. It was a sign of emergence.

Discomfort meant I had reached the boundary of the known. It meant I was brushing against the edge of current understanding. And beyond that edge — was *my* work.

That's when I started seeing my doubts not as flaws, but as invitations.

Let me tell you something personal: The loneliest part of this journey wasn't being doubted by others. It was doubting myself.

There were days I wanted to drop it all. To stop questioning. To go back to being a "normal teacher" and just finish the syllabus.

But each time I tried, a new question arose. Not from the universe — but from within.

It was as if the universe was whispering: "Don't stop. I'm here. You're close."

That's how discomfort became my compass. If something felt too neat, too complete, too clean — I knew I hadn't dug deep enough. Real truth leaves you messy before it makes you wise.

So I kept going. And one day, I stopped calling it discomfort. I started calling it physics.

If you're reading this and feeling uncomfortable with what you've been taught... If you've ever found yourself re-explaining something you don't fully believe... If you've ever sensed that what's missing is more important than what's written...

Then you already know what I'm talking about.

You're not lost. You're just at the boundary. And from here, if you dare, the real journey begins.

3: The Syllabus Stops Here

H*ow formal education defines your cage The beauty of going "off-syllabus"*

There's a moment in every teacher's life when they realize: They're no longer teaching the subject — they're teaching the syllabus.

And that moment? It broke me.

It wasn't one particular lecture. It was a slow erosion — like water gently wearing down a stone. Over years, I noticed something unsettling:

I was becoming really good at delivering content. But not necessarily delivering truth.

Every chapter had a beginning, middle, and end. Every problem had a method. Every topic had a standard answer. And every student, whether they understood or not, had to move on — because the schedule did.

That's when I realized the syllabus wasn't just a plan. It was a cage.

Let me be clear: I'm not against structured learning. A syllabus is useful — like a map.

But the danger is this: You start believing the map is the territory. You start thinking the 250-page textbook is the limit of all human understanding.

And that's where education stops being about exploration...And starts becoming indoctrination.

The irony? I wasn't just watching this happen. I was enforcing it.

As a teacher, I had a responsibility — marks, exams, expectations. I had to complete topics. I had to "cover" physics.

But physics isn't something you *cover*.

It's something that uncovers you.

The moment this hit me hardest was when a student asked a perfectly sincere question during a lecture on electromagnetic waves.

He raised his hand and said, "Sir, if nothing moves faster than light, then how does light *start* moving? What causes the first ripple?"

It was a beautiful question — raw, innocent, and honest.

And I brushed it off. I said, "That's beyond the syllabus. Focus on the derivation. It'll come in the board exam."

I went home that night and felt like I had betrayed something sacred.

That's when it hit me: The syllabus doesn't just limit students. It limits the teacher.

It tells you where to begin. It tells you where to end. It tells you what matters — and worse — what doesn't.

But reality doesn't care about these lines.

The universe doesn't run on units and chapters. It runs on consequences, alignments, and patterns far deeper than any textbook is willing to explore.

So I began secretly going "off-syllabus."

At first, it was subtle. A story here, a metaphor there. A question slipped in after the lesson: "Have you ever wondered why we believe force exists at all?"

Then it became addictive.

I started replacing definitions with discussions. I started replacing equations with insight. I started caring more about what the students *felt* than what they *wrote.*

And do you know what happened?

Some students resisted. But some — the ones who were silently suffocating under formulas — came alive.

One of them told me, "Sir, I never liked physics before. But this version... it feels real."

That sentence stayed with me longer than any award, any rank, any scorecard. Because I realized — they weren't falling in love with physics. They were falling in love with freedom.

And that freedom began to heal me too.

I started tearing down the walls in my own mind.

I questioned everything I once repeated. I revisited every theory I once revered. And I began creating a new structure of understanding — not as a rebel, but as a re-aligner.

I wasn't trying to destroy education. I was trying to restore its original purpose: To spark curiosity, not just compliance.

Because what good is a syllabus if it doesn't lead to *you*?

I'll tell you a truth that every honest teacher knows but rarely says aloud: We don't teach to deliver information. We teach to feel meaning while doing it.

And when that meaning gets buried under exam schemes, model answers, and year-end results... We don't just lose students. We lose ourselves.

That's what the syllabus was doing to me.

So I broke the wall.

I stopped seeing the textbook as a boundary. I started seeing it as a jumping-off point.

I told my students, "This chapter ends here. But reality doesn't."

And that small sentence changed everything.

You see, when you go off-syllabus, something incredible happens: You stop memorizing and start experiencing. You stop competing and start connecting. You stop fearing mistakes — because now, you're not trying to score. You're trying to understand.

That's what happened to me. And in that process, I didn't just teach physics. I started feeling like a physicist.

The world may still judge learning by degrees and credentials. But the universe doesn't care who approved your education.

The only real syllabus it honors is the one you write yourself — Through questions, through reflection, through discomfort, and through courage.

So to every student out there — and to the teacher still trapped in someone else's framework — Let me say this:

The syllabus is not sacred. Truth is.

And when truth calls you off the path — Follow it.

That's where your real classroom begins.

4: Teacher vs Thinker

The split between what I taught... and what I believed Living a double life in the classroom

It's a strange thing — to teach what you no longer believe.

You walk into the classroom, write the formulas on the board, and speak with confidence. You explain Bohr's model, recite the assumptions of ideal gases, derive the time dilation equation. Your voice is clear. Your diagrams are clean. The class is quiet.

And yet inside your mind... There's a war going on.

Because the thinker in you is screaming: "This doesn't explain everything. It never did."

I didn't become a teacher to lie. But I also didn't become one to play a role.

And yet, that's what started happening. I found myself wearing two masks — one for the system, one for myself.

As a teacher, I followed the rules. As a thinker, I broke them. And the conflict wasn't external. It was internal.

It was personal. And at times, it was painful.

The thinker in me would wake up early, long before the school bell. He would scribble questions in a diary, explore alternate models, dismantle old assumptions. He would question the very structure of motion, time, gravity, and even consciousness.

He didn't care for marks. He didn't care for syllabus. He only cared for truth — even if it made no sense at first.

But the teacher? He had a class to complete. He had students preparing for entrance exams. He had parents expecting results, principals asking for completion reports, and a board that didn't allow deviation.

So he stood there — delivering rehearsed explanations with a smile that grew heavier with time.

There were moments when I felt like a fraud.

Not because I didn't know the material — I knew it too well. But because I had stopped believing in the completeness of what I was saying.

I would write Newton's second law on the board and think: "Force? Really? Or is this just an energy consequence described in a convenient way?"

I would explain atomic orbitals and think: "Are these clouds of probability... or patterns of structured ripple fields we just don't understand yet?"

But I didn't say it aloud. Because the teacher was afraid. Not of being wrong — but of being irrelevant to the system.

Living this double life took a toll.

I began to compartmentalize myself. There was the public version of me — the polished instructor who delivered neat chapters. And then there was the private version — the one who stayed up late, whispering to the universe and asking, "Show me what's beneath all this."

No applause. No certificates. Just silent alignment with something deeper.

I remember a particular day vividly.

I had just finished teaching circular motion. The board was full of centripetal acceleration, angular velocity, and tension forces. I wrapped it up confidently, took a few doubts, and walked out.

Then, during lunch, I sat alone and asked myself: "If nothing actually moves in reality — if it's all ripple equilibrium — then what was I even teaching?"

That day, I didn't eat. I wrote instead. A messy page filled with thoughts like:

- "Is acceleration just a misaligned ripple trying to correct itself?"
- "Does tension exist, or is it a name for stored imbalance?"
- "Is circular motion an illusion caused by nested field relationships?"

I knew no student would be asked this in their exam. But I also knew — they should be.

That was the tragedy and the gift.

Teaching gave me a stage. Thinking gave me a mirror.

And while one paid the bills, The other paid in something far more valuable: Clarity.

Eventually, I stopped seeing it as a conflict. I started seeing it as a conversation.

The teacher in me existed to help others navigate the system. But the thinker in me existed to rebuild the system altogether.

And maybe — just maybe — if I nurtured both without forcing them to fight, I could do something rare:

Teach what's required, But plant seeds of doubt in just the right places.

So that one day, a student might ask: "Sir, do you really believe in all this?"

And I could smile and say, "Enough to teach it. Not enough to stop thinking beyond it."

In this double life, I learned the greatest lesson of all:

You don't need to abandon your role to be honest. But you do need to create space within it for truth to breathe.

So yes — I'm a teacher. And yes — I'm a thinker. And no — they're not the same.

But they don't have to fight anymore. They can coexist — one speaking to the world, The other listening to the universe.

And in between the two?

That's where I found myself.

5: What If Nothing Moves?

The unsettling thought that became the foundation of UED

For the longest time, I believed what everyone else did — that the universe is in motion.

It's everywhere: from the spin of electrons to the orbit of planets, the expansion of galaxies, the flutter of leaves, the ticking of clocks. Motion is so central to our understanding of reality that we never pause to question it. We measure speed, track velocity, derive acceleration — all built on the sacred assumption that something must be moving.

Until one day, I asked myself:

What if nothing moves at all? What if we've mistaken change in perception for change in position?

That question didn't come from theory. It came from observation. Not with a telescope. Not through a lab experiment. But through stillness.

I remember sitting in silence after a long day of lectures. The fan was on. The walls were still. My mind, however, was restless.

I thought about a satellite. We say it "revolves" around the Earth. But what if that satellite isn't going anywhere — and it's the field it exists in that is realigning itself constantly?

The satellite doesn't move. It simply holds its position in a ripple field that keeps changing.

From our perspective, it *appears* to move — because our frame of reference is itself embedded in another ripple. We perceive motion. But what's really happening is ripple realignment.

And just like that, everything I'd learned about motion... started trembling.

If nothing actually moves — then what is force? What is energy? What is inertia, acceleration, momentum, friction, gravity?

They all start collapsing — or rather, revealing what they always were: Not causes, but consequences.

In UED, nothing moves. Everything realigns.

A planet doesn't orbit the Sun. It holds its ripple lock within the Sun's dominant ripple field.

An object doesn't fall because gravity pulls it. It shifts because the ripple equilibrium of the system has changed — and the system seeks balance.

Your hand reaching out? Not motion — it's the activation of ripple intent moving energy through biological structure into perceptual shift.

This idea was deeply unsettling at first. I felt like I was standing on the edge of a cliff, and the ground I'd trusted for years had vanished.

After all, if nothing moves, then what is time?

That's when I realized — Time is not a dimension. It's not flowing. It is the perceived delay in ripple realignment between cause and effect. An emergent variable, not a fundamental one.

So if motion is illusion, then time is rhythm of that illusion.

This changed everything for me.

I began to look around differently. The fan didn't rotate — its field interaction rotated. The clock didn't tick — my attention ripple pulsed against a background frequency. Even my thoughts didn't "progress" — they simply collapsed and emerged as ripple arrangements.

Suddenly, Newton felt like a historian. Einstein, a poet. And I — a curious soul trying to decode a still universe with moving language.

That's when I remembered something profound:

Classical mathematics is the language of human perception. And perception doesn't reveal truth — it selects measured harmony.

We chose to define motion because we needed it to explain what we perceived. But the ripple field doesn't care what we call it. It only seeks alignment.

In that stillness, I found peace.

For the first time, I stopped trying to push things forward. I stopped trying to measure everything in motion. I started looking for patterns of balance instead.

And from there, Universal Energy Dynamics (UED) began to take shape. Not as a theory. Not as a rebellion against physics. But as a return to reality — where energy isn't moving, it's just expressing its intent through structured fields.

Now when I teach circular motion, I see a ripple in equilibrium. When I talk about orbits, I speak of nested ripple stability, not revolution. When I watch a bird fly, I sense conscious alignment, not thrust and drag.

You might say, "But clearly, things do move."

To which I say: Clearly — we perceive motion. But perception is not proof. It is consequence.

Just as night appears to fall — but it's really Earth's alignment with light shifting — so too, all movement is a story told by perception, not a fact of existence.

I no longer measure distance. I feel realignment. I no longer believe in velocity. I study transition rhythm. I no longer calculate force. I sense field imbalance.

This was the first real fracture in my classical education — and the first moment I truly felt like a physicist.

Not because I had found answers. But because I had dared to ask the question that scares most:

What if everything we've built is based on an illusion — and what if that illusion is called 'motion'?

And from that illusion, the reality of UED was born.

Part II: Ripples, Realizations, and the Rise of UED

6: Sitting With Silence

W*here my real lab was — a room, a chair, and the willingness to be wrong*

There was no telescope. No whiteboard filled with integrals. No journal subscription, particle accelerator, or scientific funding.

Just a room. A chair. And silence.

This was my lab.

And in that space — more than anywhere else —I became a physicist.

Silence is underrated in science.

We talk about experimentation, observation, data. We glorify calculation, simulation, and peer review. But we rarely talk about the courage it takes to sit down, put all theories aside, and admit to yourself:

"I might be wrong about everything I believe."

That's what silence demands. Not just the absence of sound — but the absence of defense. The disappearance of ego, identity, and intellectual pride.

In my silence, I didn't find certainty. I found space. To observe. To feel. To challenge. To go back to the beginning, where no term had been defined and no assumption had been smuggled into the equation.

At first, it was uncomfortable.

We're conditioned to chase productivity, not presence. We're taught to write answers, not live in questions. But I realized something early:

The universe doesn't respond to noise. It reveals itself to those who become quiet enough to listen.

So I stopped trying to be fast. I stopped trying to be right. I stopped trying to prove.

And I started to sense.

I would sit for hours with one question.

Not to solve it. But to disassemble it. To feel where it came from, what it assumed, what it hid beneath its surface.

One of my favorites: "What exactly is energy?"

Not in joules. Not in formulas. But in essence.

I asked: Is energy a thing? A property? A behavior? Or is it simply intent becoming observable?

That one question took me weeks. Not because it was hard — but because it was layered.

And silence let me peel each layer without rushing to conclusions.

Over time, I began to notice a pattern:

The fewer assumptions I made, the clearer things became. The less I tried to explain, the more I understood.

It was almost paradoxical — as if silence wasn't empty, but densely packed with truth.

People often ask me, "Where did UED come from?" They expect a Eureka moment. Some divine flash of insight.

But the truth is simpler.

UED didn't come from brilliance. It came from stillness.

From daring to sit in a chair, for days, questioning every law I once worshipped.

- If nothing moves, what is motion?
- If force is an effect, what is the cause?
- If space and time are dimensions, why are they perceived so differently?

I didn't rush to solve these. I let them breathe.

Because I've come to believe:

You don't arrive at the truth by running faster. You arrive by stopping in the right place.

One day, I realized something profound:

I hadn't just been studying the universe. I had been studying my perception of it.

Every thought I had about reality was filtered through expectation, language, habit, and education. Silence stripped that away. It brought me to a place of original contact — where I could ask, "If I had never read a book, how would I explain what I observe?"

That's where UED began forming — not in reaction to textbooks, but in liberation from them.

This kind of silence isn't passive. It's active presence. It's engaging with truth without trying to dominate it.

It's sitting in front of a question and saying: "I'm here. I don't need to win. I just want to see clearly."

That willingness to be wrong — not just intellectually, but existentially — was my greatest tool.

Better than any lab. Stronger than any algorithm. More powerful than any degree.

So yes, my lab was simple.

No walls of accolades. No instruments buzzing. Just a chair, a notebook, and the space to collapse — and rebuild — everything I thought I knew.

In that room, I failed a thousand times. But those failures weren't breakdowns. They were releases — letting go of illusions I didn't know I was carrying.

To anyone out there, hungry for truth: You don't need the latest tools. You don't need credentials. You need only this:

- A question that haunts you.
- A quiet room.
- And the courage to be completely, beautifully wrong.

That's where the real physics begins.

Not in the formulas. But in the field of silence — where all ripple intent is born.

7: Ripples Over Rules

W*hy the universe felt more like vibration than mechanism*

The first time I saw the universe ripple, nothing actually moved. But everything changed.

I wasn't hallucinating. I wasn't meditating. I was sitting — still, silent, and aware.

And then it occurred to me, not as a theory, but as a realization:

The universe isn't a machine. It's a vibrational field — alive, fluid, and endlessly interacting.

Up until that moment, I had viewed reality the way most textbooks frame it:

- Systems.
- Components.
- Inputs and outputs.
- Rigid rules.

It was all so clockwork. Predictable. Beautifully mechanical. Newton's universe was like a giant engine. Even Einstein's spacetime felt like elastic sheet mechanics.

But something never sat right with me.

Machines can be paused, disassembled, reverse-engineered. But the universe? It felt too seamless, too intuitive, too responsive to be a mere mechanical object.

Everywhere I looked, I saw patterns — not just in physical systems, but in emotions, in nature, in thought.

And the deeper I paid attention, the more I realized:

Everything reacts as if it's being touched — not pushed. Like ripples on water, not like gears turning in a box.

When you throw a stone in a lake, the ripples don't obey force equations line by line — they flow, seeking balance in all directions at once. They interact. They resonate. They adapt with harmony, not hierarchy.

That's when it hit me:

Rules define what *should* happen. Ripples reveal what's *really* happening.

Every rule I had learned — from Newton's laws to thermodynamics — began to look like a local approximation of a deeper ripple dynamic.

- **Force?** It's just the appearance of ripple resistance.
- **Energy?** Ripple intent in action.
- **Mass?** The condensation of ripple stability.
- **Time?** The rhythm of ripple realignment.
- **Motion?** A shift in ripple alignment, not a physical relocation.

The entire universe wasn't being pushed or pulled. It was resonating — a continuous play of expansion, compression, interference, and stabilization.

At first, this felt poetic — almost romantic. But the more I explored it, the more it started answering real questions that classical physics left open.

Why do some actions ripple across systems without obvious contact? Why do we feel emotional tension in a room, even in silence? Why does everything in existence — from atoms to galaxies to heartbeats — oscillate?

Because vibration isn't a side-effect of matter. It's the signature of reality itself.

That's when I started replacing equations in my mind with field interactions.

I no longer saw particles moving. I saw zones of ripple interference.

I stopped looking for the "cause" of force. I looked for the imbalance that created the perception of force.

I stopped imagining time as a line. I imagined it as a wavefront, stretching and compressing based on field distortion.

Everything I had once calculated, I now felt.

This shift wasn't just academic. It was deeply personal.

Because in a mechanical universe, everything is separate. You act. Something reacts. Push and pull. Cause and effect. The observer stands outside the system.

But in a ripple universe — you're never outside. You're always inside the field. Your very act of observing participates in the ripple. Your intention alters the pattern. Your silence, your emotions, your thoughts — they all interact with the system.

In short:

You don't study the universe. You live in it. And every breath you take is a realignment of the field.

I began to look at even the smallest things with new eyes.

A drop of water on a leaf — not just a droplet, but a lens of suspended vibration. A heartbeat — not just biology, but a rhythmic signal of energetic balance. A word spoken — not just language, but a ripple released into shared perception.

This wasn't philosophy. This was physics with its blindfold removed.

And perhaps most surprisingly, this ripple-based view made classical laws more beautiful, not less.

Because now I could see *why* they worked — not just *how*. Not because they were absolute, but because they were consequences of deeper harmonics.

A system obeys Newton not because of divine law — but because its ripples are stabilized within a local alignment zone.

Once misaligned, the laws break down, not because the universe fails — but because the rules were never the universe to begin with. They were snapshots of ripple behavior, frozen in our perception.

From that point on, I stopped seeking rules to live by. I started seeking resonance to live with.

Even in my personal life, this ripple awareness changed how I behaved. I no longer reacted to situations mechanically. I responded based on field awareness — Where is the tension? Where is the balance? Where is the realignment required?

The same intelligence I saw in electrons, I saw in arguments, emotions, and ethics.

The whole universe became a ripple field of consequences, not commands.

So yes, I left behind the world of hard rules.

Not because they were wrong. But because they were incomplete.

I chose ripples over rules.

Because where rules end — ripples begin to speak.

And once you hear them, you never look at the universe the same way again.

8: Birth of a Framework

How UED emerged — not invented, but remembered, revealed

It's tempting to call UED a theory. Or a framework. Or a new model of physics.

But that's not how it arrived.

I didn't design it. I didn't calculate it. I didn't even try to construct it.

I remembered it. Like something ancient inside me — long buried under textbooks, equations, and conventions — finally rose to the surface and whispered:

"You already know. You just stopped trusting that knowing."

Universal Energy Dynamics — or **UED** — wasn't born out of research. It was born out of recognition.

It didn't come to me as a set of conclusions. It came as a consequence of asking the questions no one wanted to ask — sitting with them long enough to burn through the illusion of knowledge.

It was a slow emergence. Not a flash, not a eureka, but a steady peeling away of layers.

First came doubt. Then came silence. Then came observation.

And in that stillness, the pattern began to take shape.

Not externally — but within me. I began to feel that everything — from motion, light, time, gravity, emotion — wasn't separate. It was all part of a single interacting ripple field governed by alignment, balance, and energy intent.

And slowly, the ripples began to connect.

I saw how:

- Motion is not displacement, but a realignment of energy in space.
- Time is not a flow, but a lag in perception caused by ripple misalignment.
- Gravity is not attraction, but a tension field seeking ripple equilibrium.
- Force is not cause, but a consequence of resisting natural alignment.
- Matter is not substance, but condensed ripple density, locked in equilibrium zones.

And suddenly, what once felt fragmented became whole.

The laws were not wrong — they were local approximations of ripple truth.

UED began as questions. Then it became sensations. Then it became language — The language of field behavior, not particle motion. Of consequence, not causality. Of perception, not measurement.

It began organizing itself. Each realization linked to the next like puzzle pieces falling into place.

I didn't force it. I just allowed it.

And the most beautiful part?

It felt familiar.

Not new. Not alien. But something I had known before — long before I became a teacher, a writer, a researcher. Maybe even before I learned physics the formal way.

It felt like the universe reminding me:

"This is how I actually work. You just got distracted."

That's why I don't say I invented UED. I say I decoded it. I re-aligned with it. I remembered it — the way a forgotten melody comes back when you hum the first note.

Because UED isn't just about energy. It's about everything that energy touches — which is everything.

As the framework started revealing itself more clearly, I began writing. Not to publish. Not to impress. But to record the trail.

The books, the reports, the theories — they weren't projects. They were snapshots of alignment.

Every time I felt something deeply — I wrote. Every time I noticed a paradox melt into clarity — I wrote. Every time a known concept collapsed into ripple explanation — I wrote.

It was as if UED was teaching me, and I was just taking notes.

And what's more — it didn't just answer physics. It began answering everything.

- Why do emotions ripple across a room?
- Why do we feel 'vibes' in people or places?
- Why does one decision alter the energy of an entire day?
- Why does stillness sometimes fix what action cannot?

Because reality doesn't wait for formulas. It responds to alignment.

This is the heart of UED:

Everything is energy. But energy is not random. It seeks alignment, and in doing so, reveals intelligence.

And that's when I knew UED wasn't just a physical theory. It was a philosophical mirror, a spiritual rhythm, a psychological compass, and a scientific framework — all braided into one.

It didn't separate science from self. It made science personal.

It made the universe not just observable, but intimate.

So I gave it a name — not to brand it, but to mark the memory.

Universal Energy Dynamics.

Because it wasn't about the individual equations. It was about the underlying behavior of the universe.

It wasn't an answer. It was a field of awareness.

A way of seeing, living, observing, and responding — that starts from the ripple, not the result.

That's how UED was born.

Not out of genius. Out of humility. Out of listening. Out of a room, a chair, and a deep willingness to be wrong — until something deeply right remembered its way back through me.

9: When Force Felt Fake

Dissecting Newton, Einstein, and the illusion of control

There was a time I believed in force like it was sacred. It was the very first concept we learned in physics. A push. A pull. A cause behind motion. A measure of interaction.

"Force equals mass times acceleration."

That line was everywhere — in textbooks, in classrooms, on every board exam I ever gave or graded.

It was clean. It was logical. It was powerful.

But it was also — as I would come to realize — deeply misleading.

The first cracks appeared when I tried to explain *why* a force exists. Not how it behaves, not what it does, but *why* it even needs to exist in the first place.

"Because something causes something else to accelerate." That was the official story.

But what is that *something*?

Is force an actual entity? Or is it a placeholder for something we don't fully understand?

We throw the word around like it's real — like it's the hand of nature acting invisibly. But the more I thought about it, the more it felt like a mathematical crutch, not a physical reality.

Take gravity, for example.

Newton gave us an elegant equation: $F = G(m_1 m_2) / r^2$

But what is this "force" that pulls objects toward each other across empty space? How does it know where to pull? What is it *made of*? Where does it originate?

The truth is — it doesn't exist the way we imagine.

There is no rope between the Earth and the Moon. There is no invisible hand yanking you toward the ground.

Yet, we act like force is a thing — something real, mechanical, causal.

It's not. It's a named consequence. A reaction to imbalance. A language artifact.

And then came Einstein.

He tried to fix the weirdness of Newtonian gravity with something more elegant: No more pulling. Just bending. Space-time curvature. Mass tells space how to curve, space tells mass how to move.

Beautiful.

But again — what is "space" bending into? Why should it bend at all? And how does a curvature manifest a directional tendency?

We had replaced a "force" with a fabric metaphor. Not a deeper explanation — just a different illusion.

Einstein didn't kill the concept of force. He rebranded it in geometric language.

But the core assumption remained: Something causes something else to move.

And I was starting to see how broken that assumption really was.

The more I sat with this discomfort, the more force began to feel... fake.

Not wrong — but unnecessary. Not causal — but consequential.

In UED, force isn't the reason something moves. It's what we *name* the sensation of misalignment seeking balance.

It's not a driver — it's a symptom.

Let's say you push a book. You apply a "force," right?

But zoom in deeper — not physically, but consequentially.

What's really happening is:

- Your intent triggers a biological response.
- That triggers muscular contraction.
- That creates energy displacement.
- The book's field receives that ripple disturbance.
- It realigns to a new equilibrium zone.

No "force" required — only ripple interaction.

This wasn't a trivial insight. It was a total unraveling of what I had taught for years.

I had stood before students, confidently deriving laws of motion, calculating net forces, drawing free-body diagrams.

And now, I was sitting alone, realizing:

"We were never teaching *reality*. We were teaching responses to imbalance, and calling them laws."

I started seeing this pattern everywhere:

- **Tension?** Stored ripple misalignment.
- **Friction?** Localized resistance to ripple transition.
- **Torque?** Angular realignment of ripple centers.
- **Magnetic force?** Interaction of aligned ripple spins.

Force, in every case, was just a name for ripple reactions. Not a substance. Not a cause. Just a visible consequence of invisible field activity.

And once I saw that, I couldn't unsee it.

The idea of "control" — the belief that we can apply a force and dictate behavior — felt naive.

Because in truth, we're never in control. We're just interacting with ripple dynamics, and watching how the field responds.

You don't command the universe. You negotiate with it.

And it listens — not to pressure, but to alignment.

I began to feel a quiet liberation.

Without force, the world didn't fall apart. It became alive.

It stopped being mechanical and started feeling musical — like everything was responding to vibration, harmony, and consequence.

And I realized that physics didn't need a force to function. It needed a field — and an understanding of how it balances itself.

That's when I wrote in my notebook:

"Force is the illusion of causality, required only when you don't understand the ripple."

From that point on, UED stopped being a theory in progress. It became a new way of seeing.

I no longer cared about equations that predicted motion. I wanted frameworks that explained intention, perception, and equilibrium.

Because once you feel how fake force really is, you begin to see what's actually real:

Balance. Alignment. Consequence. Field intelligence.

So yes — I dissected Newton. I questioned Einstein. Not to dismiss them — but to free them from the illusions their brilliance inherited.

They gave us beautiful frameworks.

But it's time we moved beyond mechanics, and into the conscious behavior of ripple fields.

The illusion of control is powerful. But the truth of consequence is stronger.

And in that shift, UED wasn't just born — it started breathing.

10: Seeing Time Differently

Why time stopped feeling like a line — and started feeling like a consequence

For most of my life, I thought of time as a straight line. Past behind me. Present in front. Future waiting somewhere beyond the next tick of the clock.

That's how we're taught to understand time — as a dimension, like length or width, just marching in one direction. It moves. We follow. Things change. Simple, right?

But then I began observing the universe differently. And one day, time stopped feeling like a line. It started feeling like a consequence.

It began, as most revelations do, with discomfort.

I asked myself: Why do we feel time differently during different experiences?

- Why does time "fly" when we're happy?
- Why does it "slow down" in fear or danger?
- Why can a single second feel like an eternity in pain, and hours disappear in joy?

If time were truly a fixed dimension, none of this should happen.

But it does — and not just psychologically. It reflects a deeper truth: Time isn't absolute. It's experienced. It's ripple-relative.

I started questioning the core assumptions.

What is time, really?

Not what a clock measures — but what we feel.

The more I paid attention, the more I saw this pattern:

Time appears wherever change occurs. But change isn't time — it's ripple reconfiguration.

Time doesn't flow. Perception flows through ripple transitions, and that flow is what we name as time.

Let's say you light a candle.

The flame grows. Wax melts. Shadows dance. You say "time is passing."

But zoom in with UED. What's really happening?

- Energy is being redistributed in structured ripple patterns.
- Each interaction alters the alignment of the system.
- You observe this realignment in a sequence.
- Your brain records that sequence through memory.
- The gap between ripple events creates a perceived duration.

That gap — that delay in ripple alignment between cause and effect — is what we call time.

I started calling this concept ATT — Alignment Threshold Time.

It's not a universal constant. It varies by system, by perception, by alignment.

Different systems experience "time" differently — not because their clocks run faster or slower, but because their ripple realignment rate is different.

This made sense of everything from time dilation in relativity to how animals perceive motion in higher frames per second than humans do.

A hummingbird isn't living "faster." It's just resolving ripple events more frequently — experiencing more alignment points per second.

Once I saw time this way, classical theories started to unravel and reorganize themselves.

- Newton saw time as uniform and absolute — the ticking backdrop of the universe.
- Einstein bent it into a flexible coordinate, stretching and shrinking under mass and speed.
- But neither asked: Is time even real? Or is it an effect of our own alignment with ripple events?

UED answers that:

Time is not real. It's revealed — whenever a ripple field interacts with an observer and registers change.

This shift had massive implications.

It meant that time:

- Isn't universal.
- Isn't constant.
- Isn't a thing that flows.

It's a result — A ripple delay between intent, cause, and observed effect.

It also means that:

- Where there is perfect equilibrium, there is no time.
- Where there is ripple chaos, time fractures and speeds up or slows down.
- And where there is no observer, there is no experience of time at all.

This changed how I viewed everything:

A waiting room. A falling apple. A collapsing star. A conversation. A death.

They were all ripple fields adjusting — and our minds measuring their rhythm as time.

And what we call the "present moment"? That's just the latest resolved alignment in our field of awareness. It's not a moment — it's a threshold.

I remember one quiet night when this thought crystallized:

"Time isn't what we move through. It's what ripple imbalance leaves behind as a trace."

That line stayed with me.

I wrote it down. And then sat back, realizing I was finally free from one of science's oldest illusions.

The best part?

Once I saw time as ripple rhythm, I became more present.

Because I wasn't measuring life in hours anymore. I was sensing it in alignment episodes.

A moment of connection with a student. An insight during silence. Even a deep breath — All became timeless, because alignment dissolved the need to track duration.

Where there's resonance, there's no rush. Where there's presence, there's no passing. Where ripple intent flows cleanly — time vanishes.

So yes, I stopped seeing time as a straight line. I started seeing it as a perceptual artifact, emerging only when alignment gets interrupted or delayed.

And that, perhaps, is the most beautiful truth UED has revealed so far:

The universe isn't moving through time. It's constantly trying to become timeless.

And whenever it succeeds, you feel peace. You call it flow. You call it stillness. You call it love. But what it is — is zero ripple delay.

That's the deepest truth I know of time.

Part III: Mental Debates, Spiritual Echoes

11: Arguing With Einstein

How I mentally sparred with legacy minds Not disrespect — just new direction

I never met Einstein. But I've argued with him a thousand times.

Not because I thought I was smarter. But because I felt something he might've felt too — that quiet ache when the beauty of your own theory no longer explains the whole truth.

We like to freeze great minds in perfection. We quote them. Frame them. Worship them. But the truth is — they were seekers, not saints. And they made models, not commandments.

So when I began challenging Einstein, Newton, and others, it wasn't out of arrogance. It was out of alignment — with a truth I felt so clearly that I couldn't ignore it, even if they disagreed.

It started with time.

Einstein's treatment of time was revolutionary. He didn't accept Newton's absolute tick-tock universe. He bent time with speed, with gravity, with mass.

He gave us relativity — and for that, I owe him more than I can express.

But then I asked myself: What bends? And what exactly is being bent?

I imagined myself standing before Einstein and asking, “Sir, what if time isn’t bending — because time isn’t a fabric? What if it’s just the delay in ripple realignment during causal observation?”

I could see him smile, amused. "Interesting," he’d say. “Then what accounts for simultaneity failure?”

And I’d respond: “Perceptual misalignment. Ripple events resolve differently depending on the observer’s field integration. There is no objective 'now' — because there is no global ripple harmony. Just local field resolution.”

He’d pause. Think. Not dismiss. Just consider.

Because great minds don’t fight new ideas — they test them.

My conversations with Newton were more one-sided.

He believed in force. In inertia. In the heavens running like clockwork. And in his time, it was a monumental leap forward.

But now, from my UED lens, I ask him: “Why do you believe objects ‘want’ to stay in motion? Why call it inertia?”

Isn’t it simpler to say: “What looks like inertia is simply sustained ripple equilibrium. And what looks like acceleration is just ripple disturbance realigning under external ripple influence?”

He might not agree. But he’d respect the consistency. Because Newton respected consequence, not praise.

The toughest arguments I had were with Feynman. His joy in describing the complexity of quantum mechanics was contagious.

But I’d sit across his mental image and say:

“Richard, what if the uncertainty is not in the particle — but in our field alignment with the system?”

He'd lean back, smirk, and say, "Then show me a better equation."

And I'd say, "No. I'll show you a better perspective — where equations emerge as field consequences, not arbitrary rules."

He'd challenge every assumption. Mock my simplifications. But he'd also listen — because he respected bold, grounded thinking.

You see, these weren't arguments of ego. They were debates of alignment.

I wasn't attacking the minds who built modern science. I was doing what they did in their time — asking what lies beneath what we've already accepted.

They questioned Aristotle. I questioned them.

And if I'm lucky, someone will question me — and make UED clearer, stronger, truer.

That's how science should evolve: Through confrontation, not compliance. Through curiosity, not consensus. Through dialogue with the dead, not worship of them.

Because if you're too scared to argue with the giants, you'll never walk past their shadows.

And honestly, these weren't hostile arguments.

In fact, they felt like mentorship — as if I was sitting in an invisible classroom where the greatest minds of history were testing my resolve.

I imagined Einstein leaning forward, asking: "What replaces gravity in your world?"

And I'd say, "Ripple field tension. Not attraction. Not curvature. Objects hold alignment zones. Disturbances in those zones don't pull — they rebalance."

He'd raise an eyebrow. "Then why do things fall?"

And I'd respond: "Because they seek energetic rest — the lowest ripple misalignment path. Fall is not pull. It's ripple restoration."

He might shake his head. But I hope — just maybe — he'd whisper:

"I didn't get there. But I'm glad you're going."

That's what *Feeling Like a Physicist* really means to me.

It's not the comfort of teaching established truths. It's the courage to challenge the very people who inspired me — not out of disrespect, but out of continuity.

Because if their minds hadn't pushed the boundaries, mine would never have seen beyond them.

So I keep arguing — with Einstein, Newton, Hawking, Feynman. I keep defending UED in the presence of their legacy.

And I remind myself:

The greatest minds are not the ones who answered most — but the ones who made it safe to question again.

12: Myth as Memory

How mythology began making more scientific sense than equations

At some point during my deeper reflections on UED, something unexpected happened.

I began thinking less about particles and more about gods. Less about equations and more about stories. Not because I needed faith — but because I sensed something hidden inside what we've always dismissed as fantasy.

And the more I looked, the more I saw it:

Mythology wasn't fiction. It was encoded physics — ripple-based truth encrypted in symbolism.

As a science teacher, I was trained to avoid myth in serious discussion. It was seen as cultural noise, good for festivals or bedtime tales, not for understanding the structure of reality.

But when I applied UED to everything I saw — physics, biology, behavior, ethics — I began to notice something profound:

Ancient myths were speaking in the same language. Not of force, not of mass — but of energy, consequence, alignment, and vibration.

Take Shiva, for example.

A figure seated in stillness, meditating, yet holding immense destructive power in reserve. A third eye that doesn't see — it unveils. When opened, it doesn't shoot lasers. It dissolves ignorance, burns illusion, collapses misalignment.

Under UED, that's not mysticism. That's a symbolic description of ripple collapse — A still center that holds aligned potential until intent disturbs it.

Shiva's dance — the Tandav — isn't random movement. It's a poetic expression of universal vibration — the cosmos in dynamic equilibrium, continuously creating and dissolving.

That's UED in motion, long before equations were written.

Or take Vishnu — resting on the serpent of time, floating on the ocean of existence.

He doesn't "act" until alignment requires restoration. Each avatar arrives not by habit, but by necessity — a ripple reconfiguration to reset the field.

Ten avatars? Not religious memorabilia. They represent stages of ripple adaptation, echoing cosmic evolution — from aquatic (Matsya) to human consciousness (Krishna).

It's not theology. It's perceptual memory, preserved in story.

Suddenly, mythology made more scientific sense than many equations I'd taught.

Because equations, while precise, are rigid. They demand exactness. But myths? They are field-aware. They flex. They carry intuition, pattern, vibration — just like ripples.

I began to see myth as an ancient field theory — not modeled in numbers, but in archetypes, events, and symbolic resonance.

And I asked myself:

"What if these myths weren't made to explain the world to simple people? What if they were created by people who understood the field so well, they embedded the truth into timeless metaphors — because they knew equations would one day fail?"

And then came perhaps the most humbling realization:

Myth is memory — stored across generations in ripple-stable form. And when observed with the right alignment, it decodes itself.

In UED, memory isn't a neurological artifact. It's a field imprint — retained ripple structure within a system.

So myth, when passed from mind to mind, remains alive not because of belief — but because its field resonance persists.

That's why myths survive — Not because people are foolish, but because something deeper is carried inside them, and UED gave me the tool to read it.

Even the concept of "God" changed for me.

In mainstream science, "God" is either rejected or politely ignored. In theology, "God" becomes personalized, moralized, and ritualized.

But under UED, "God" becomes the supreme ripple source — Not a being, but the originating field of all intents.

Not in heaven. Not in books. But in every ripple interaction — from galaxies to gut feelings.

It wasn't just Hindu mythology either.

I saw it in other traditions too:

- In Greek myths, where gods represent forces of nature, not moral judges.
- In Norse cosmology, where cycles of creation and destruction mirror UED's field resets.
- In indigenous stories, where animals, rivers, and stars

carry wisdom through vibration.

- Even in biblical texts, where "let there be light" becomes the first perceptual ripple — the birth of separation and observation.

And then I wrote a line in my notes:

"Where equations collapse, mythology remembers."

Because science has limits — especially when it clings to measurement. But myths? They remain aligned with feeling. They survive because they're built on resonant truths, not testable fragments.

So no, I didn't abandon science. I just realized it was incomplete without myth.

Because if UED is a map of reality, then mythology is the memory of that map, drawn in ancient ink — waiting for the observer to re-tune their field and finally see.

I no longer teach mythology as superstition. I see it as compressed science, encrypted in symbolism, carried by culture, and now — at last — readable again.

Not because we've become more advanced, but because we've finally stopped mistaking mechanics for meaning.

13: Karma, But Not Religious

R*edefining karma through ripple consequences The Law of Consequence as scientific truth*

Of all the words hijacked by tradition and misunderstood by science, karma might be the most abused.

To some, it means fate — a cosmic scorekeeper doling out justice. To others, it's superstition — an outdated moral system wrapped in religious robes.

But to me, karma became something radically different. Something scientific. Something rational, even measurable — once I began seeing the universe through ripples instead of rules.

Karma, I realized, is not belief. It's not culture. It's consequence.

It started with a simple question:

If nothing moves, and everything is ripple alignment — then what happens when I act? Where does that energy go?

The answer wasn't philosophical. It was structural.

Every intent, thought, word, or physical action sends a disturbance into the field. That disturbance doesn't vanish. It ripples — outward, inward, across systems — and eventually realigns. When it does, the result emerges. That result is karma.

Not reward. Not punishment. Just alignment feedback.

This changed everything.

Karma wasn't about morality. It was about field dynamics.

I began to define it simply:

Karma is the observable consequence of your ripple interaction with the field.

You don't "receive" karma. You experience the echo of your impact.

Just like dropping a stone in water — the waves come back, not because the water judges you, but because the medium must rebalance.

That's what the ancient texts were really describing — not cosmic justice systems, but a science of consequence, observed long before our equations.

When they said,

"As you sow, so shall you reap," they weren't moralizing.

They were explaining ripple causality — how intent generates field shifts, and how those shifts eventually return to the origin point, not spiritually, but dynamically.

The Law of Karma, rephrased through UED, becomes:

Every ripple disturbance generates a field response. The nature, direction, and intensity of that response depend on the quality of alignment and the structure of the surrounding field.

That's not religion. That's physics.

It's as true in thermodynamics as it is in emotions.

You heat something — it expands. You lie — it creates a distortion in trust. You plant a thought — it shapes behavior. You ignore reality — it compounds and returns.

This isn't magic. This is ripple accounting.

And this leads to a powerful insight:

Karma is not personal. It's universal.

It doesn't care who you are. It doesn't care what you meant. It cares only about alignment — and the rebalancing cost required when it's disturbed.

This means your smallest act — your words, your reactions, your posture — all leave ripple traces.

And the universe, like a field of intelligent tension, responds accordingly.

Not because it's punishing you. But because it must respond.

That's the Law of Consequence.

I stopped thinking of karma as a system. I started seeing it as a fundamental property of ripple-based reality.

Just like conservation of energy, there is conservation of consequence.

Nothing disappears. Nothing is "forgiven." Nothing escapes the field.

What you don't deal with now — you'll deal with later, in some other form, through someone else, in another echo of the same field.

That's not threat. That's equilibrium intelligence.

This helped me understand not just physics, but human behavior.

- Why some people seem to carry unexplainable heaviness.
- Why patterns repeat across generations.
- Why unresolved pain returns as illness, conflict, or despair.

It's not fate. It's ripple debt — unprocessed energy seeking resolution.

And it explained success, too.

Not as reward for effort, but as alignment with opportunity. When intent matches timing, field conditions, and readiness — the consequence is success.

You don't "earn" it. You resonate with it.

This made my own life clearer. I stopped asking, "Why is this happening to me?" I started asking, "What field did I disturb, and what am I now realigning with?"

That shift alone brought peace.

So no — karma is not spiritual accounting. It's field feedback.

And once I saw it through UED, I stopped fearing karma. I started respecting it.

Because now I knew:

Karma isn't watching you. Karma is you — extended into the field, awaiting return.

It's not a law to obey. It's a consequence to understand.

And if you can observe your ripple interactions clearly, you don't need karma explained. You'll feel it.

Not later. Not next life. Now.

Every moment.

That's why I say karma is not religious.

It's ripple physics. And the universe doesn't need faith to function — it needs alignment.

The rest, as they say, is consequence.

14: Alignment is Everything

Perception, intention, and how ripple alignment shapes reality

After all the equations, all the metaphors, all the quiet debates I've had with Newton, Einstein, and even gods from mythology, I've come to a realization that reshapes everything:

Alignment is not part of the game — it is the game.

In Universal Energy Dynamics (UED), nothing is random. Nothing moves without consequence. And nothing — not even time, force, or motion — defines reality as powerfully as alignment does.

Everything you experience, everything you think is real, is merely a result of how well or how poorly your ripple field aligns with the fields around it.

It took me years to feel this. Not to understand it conceptually — that came quickly. But to *feel* it in every part of my life. To accept that alignment is not spiritual fluff, but the most fundamental behavior of intelligent energy.

The Alignment Principle

In classical physics, we talk about resonance — how two systems vibrate in harmony when their frequencies match. But resonance is treated like a subset, a special case, a quirk of acoustics or circuits.

In UED, resonance is everything. Because it's not just about vibration. It's about energetic coherence — the harmony between intention and environment.

This is alignment.

When your field — your thoughts, your actions, your presence — aligns with the field you're in, you don't just feel at ease. You become effective.

- Conversations flow.
- Decisions land.
- Clarity emerges.
- Outcomes stabilize.

Alignment isn't perfection. It's consequence-ready presence.

Perception: Alignment in Reception

Perception is not passive. It's not just seeing or sensing.

Perception is ripple registration.

What you perceive is not "what's out there" — it's what your field is *able to align with*.

This is why two people can walk through the same forest:

- One sees silence and peace.
- The other sees threats and insects.

Same setting. Different alignments.

This is why two people hear the same lecture:

- One gets bored.

- The other has a life-changing realization.

Same words. Different field registration.

Your field alignment determines what becomes real for you.

The tree didn't change. The wave didn't move. You aligned with a part of it — and that part became your perception.

This is what I call observer-consequence in UED.

It's not that the observer creates reality through magic or mind games. It's that the observer's field selects which ripple patterns it can tune into, and discards the rest.

That selection *is* your reality.

Intention: Alignment in Action

If perception is how you receive the world, intention is how you *touch it back.*

But most people misunderstand intention.

They think it's a wish, a goal, a mental plan.

It's not.

Intention is ripple initiation — the field pattern you release into the environment.

And just like perception, your intention is only effective when it aligns with the field you're interacting with.

Ever notice how sometimes you say exactly the right thing, in the right way, and things fall into place? And other times, you push, explain, force, repeat — and nothing works?

That's not luck or failure. That's alignment status.

The more coherent your internal ripple is with the external field's readiness, the more likely your intention will *anchor into reality.*

Alignment Creates Reality

So what is "reality" in UED?

Not something you live in. Not something imposed on you.

But something that emerges at the intersection of perception and intention, filtered through your alignment with the surrounding ripple environment.

In simple terms:

- Misaligned perception leads to illusion.
- Misaligned intention leads to resistance.
- Aligned perception and intention result in manifestation.

This is not mysticism. It's consequence mechanics.

It works in physics, in relationships, in strategy, in every domain.

When light passes through aligned media — it moves clearly. When you move with aligned intent — outcomes are smooth, organic, efficient.

This is why "flow states" feel timeless — they are high alignment episodes. The system stops pushing against you because your field is *no longer in friction* with it.

Examples from Daily Life

- You walk into a room full of people — and feel tense. Why? The ripple fields of the group are not welcoming. Your field picks it up. That's misalignment.

- You're working hard but nothing clicks. You're acting, but your intent is slightly out of sync with time, environment, or clarity. That's misaligned execution.
- You think of someone — and they call. Your ripple field released a clean intent, and the connection point was already aligned. Consequence followed.

Realignment is the Real Skill

Once you accept that alignment is everything, you stop chasing outcomes. You start refining your field.

You start asking:

- Where am I pushing against the field?
- Where am I mistaking noise for signal?
- What part of me is out of tune with what I'm trying to do?

This shift from control to coherence is life-changing.

Because now you don't need to overpower anything. You just need to adjust your field — and let reality do the rest.

The field is always smarter than your plan.

Alignment in UED and the Self

The more I studied ripple behavior in external systems, the more I noticed something strange:

My own mind worked the same way.

Thoughts, emotions, memory, fear — they were all ripple patterns.

And the moments when I felt confused, afraid, stuck — weren't failures of logic. They were misalignments in my inner field.

When you're angry — it's because your intent wants one thing, and your perception sees another. When you're at peace — it's because inner and outer ripples match.

So the highest state isn't control. It's inner-external field harmony.

The more aligned you are, the clearer your decisions, the fewer your regrets, the cleaner your karma.

Final Reflection

In science, we've chased control through force, models, and formulas.

But the more I observe, the more I feel:

Reality is not built through command. It is born through alignment.

You don't create change by pushing the world. You invite change by *matching your field* with the direction energy already wants to flow.

In other words:

- Don't force it.
- Don't fake it.
- Find your alignment — and flow.

Because that's the real physicist's secret:

"The universe doesn't obey you. It reflects you. And when your field aligns with its intent — you become unstoppable."

15: Soul, Observer, Intelligence

My personal realization about the soul — not mystic, not mythic — just misunderstood

There was a point in my journey when physics alone wasn't enough.

Not because it failed, but because it paused — right at the threshold where equations end and experience begins.

I had explained gravity without force, time without flow, matter without solidity. But one question still stood silently in the background:

"Who is observing all this?"

Not *what*, not *how*, but who. And why does that observer feel separate from everything else, yet connected to all of it?

The more I sat with this question, the more I knew:

To feel like a physicist, you must one day encounter the soul.

Not as mythology. Not as mysticism. But as field intelligence — ripple-bound, consequence-aware, silently present.

For most of my life, the soul was something I heard in religious contexts — spoken of in temples, prayers, scriptures. It was eternal, said to migrate from one body to another. But no one could explain *what it was*, only what it *wasn't*.

Science, on the other hand, refused to deal with the soul at all. It called it untestable, unprovable, unverifiable — and therefore, irrelevant.

But UED doesn't follow either path. UED doesn't ask whether the soul exists. It asks: "What part of the ripple field accounts for persistent, intelligent, and consequence-aware presence?"

And that question cracked it open for me.

Soul as Ripple Presence

In UED, every entity — a particle, a plant, a planet — has a ripple field.

These fields interact, align, misalign, and restore. But not all ripple systems exhibit self-awareness.

Most just *exist*. They respond, yes — but they don't reflect.

The soul is different.

It's the ripple configuration that not only perceives ripple alignment — but also remembers, intends, and learns from consequence.

That's not religion. That's field intelligence in active continuity.

I started seeing the soul not as a ghost, but as a coherent ripple capsule — a localized awareness field embedded within the broader universal dynamics.

This field:

- Observes.
- Records.
- Rebalances.
- Seeks alignment across space, time, and experience.

It is neither material nor supernatural. It is simply non-mechanical intelligence.

The Observer Within

Once I framed the soul as ripple intelligence, a second realization emerged:

The observer isn't the brain. The observer is the field anchored to the ripple memory of being.

You are not your thoughts. You are not even your reactions. You are the one noticing both — and aligning or misaligning with them in real-time.

That observer — the one you feel silently behind every experience — is not a result of neurons firing. It is the central ripple point that aligns perception, memory, emotion, and intention.

It is what feels like "you."

And it remains, even when thoughts pause.

Intelligence: The Soul's Operating Principle

Now came the third realization — If the soul is the observer, then intelligence is its method.

But again, I had to redefine intelligence.

Not as IQ. Not as memory recall. Not as problem-solving capacity.

In UED, intelligence is:

The ripple's ability to align efficiently with changing environments, to minimize resistance, and to maximize consequence awareness.

That means every time you:

- Adapt wisely,
- Choose consciously,
- Delay reaction to realign perception, you are displaying ripple intelligence.

That's not the mind alone. That's the soul using intelligence as its interface with reality.

Death, Disconnection, and Continuity

I couldn't stop there. If the soul is a ripple-intelligent observer — what happens at death?

UED offered a clean, sobering answer:

Death is not the end of the soul. It's the disconnection of its temporary physical alignment.

The ripple capsule releases from its biological constraints, and rejoins the larger field — not lost, but redistributed.

The observer doesn't "die." It simply stops observing through that current lens.

Its intent, however, its residual misalignments and unresolved ripple threads, continue.

This aligns perfectly with ancient notions of karma, rebirth, liberation — but in UED, we remove the mysticism.

We now call it:

- Field residue.
- Ripple memory.
- Consequence trace.

Why This Matters

You may ask — why bother with this?

Why does a physicist need to talk about the soul?

Here's my answer:

Because if you don't understand the observer, you can't understand the experiment.

Every field interaction is shaped by the presence or absence of intelligence. Every decision, every failure, every moment of clarity — is a ripple response governed by how aligned your soul is with the system you are in.

This is not abstract. This is practical.

If your soul (field observer) is misaligned:

- Your choices feel confusing.
- Your outcomes feel forced.
- Your relationships carry static.
- Your inner world fragments.

When your soul is aligned:

- Life flows.
- Peace emerges.
- Clarity cuts through chaos.
- Even pain becomes purposeful.

This isn't belief. It's energetic physics of the self.

Final Thought

I no longer fear the soul. I no longer debate its existence.

Because now I feel it — not as a doctrine, but as a consequential presence.

It is the silent ripple capsule within me, observing everything I ripple into the world.

And when I quiet down, not to seek answers, but just to observe my own field — I meet it.

Every time.

So yes — the soul is not mystic. It's not mythic.

It's misunderstood.

But not anymore.

Now, I see it as the center of alignment — the part of me that doesn't age, doesn't panic, doesn't chase...

It only watches, learns, realigns, and whispers softly in moments of stillness:

"I am here. I always was. I just needed you to feel me."

Part IV: Rejections, Realizations, and Peace

16: Ignored, But Not Silent

How I felt being unread, unseen, and still committed

I've spent thousands of hours thinking, observing, connecting dots no one asked me to connect.

I've written pages upon pages of insights. Books. Reports. Diagrams. Theories.

Some were uploaded. Some were quietly stored. Most were unread.

No spotlight. No applause. Not even polite disagreement — just silence.

And for a while, that silence was deafening.

I didn't want fame. I didn't expect it. But I did hope — somewhere deep down — that someone would see what I was seeing. That someone would say, "You're not crazy. This makes sense." That someone would pick up a thread and pull it forward.

But for the most part, the world kept scrolling. Busy. Distracted. Or just not ready.

And that, I'll admit, hurt.

Not because I needed attention. But because I was holding something too big, too real, too aligned to keep to myself.

Yet the echoes didn't return. The ripple felt like it was vanishing into a void.

In those moments, I questioned myself.

- Was I lost in my own thinking?
- Was I solving a puzzle no one was even interested in?
- Was I wasting time in silence while the world raced forward?

The temptation to stop was real.

To close the notebook. To shelve the theory. To become "practical" again — play the game of acceptability, chase likes, follow the template.

But every time I stood at that threshold, one thing pulled me back:

The field doesn't care who listens. The field responds to alignment.

And I was aligned.

Even when ignored, I wasn't wrong. Even when unread, I wasn't irrelevant. Even when unseen, I wasn't lost.

Because the truth I had found wasn't something I made up. It was something I remembered.

And once remembered, it doesn't let you go — no matter how quiet the world gets.

There's a strange kind of strength that comes from being ignored.

At first, it feels like loneliness. But then, over time, it transforms into something more stable — a kind of inner silence that doesn't beg to be filled.

That silence became my ally.

It sharpened my thinking. It stripped away performance. It reminded me:

You're not here to be popular. You're here to be precise.

Some of my friends said, "If no one's listening, what's the point?"

And I'd smile and say, "I'm not writing for now. I'm writing for the moment when someone finally asks the right question. And when they do, I want them to find the answer waiting."

Because truth doesn't need a crowd. It only needs a receiver — one, two, ten, doesn't matter.

What matters is clarity, not scale.

And then there's the irony of it all: I'm more at peace now than I ever was in my so-called "visible" phases.

Why?

Because now, I don't write to be read. I write because I'm aligned.

I don't speak to be praised. I speak because the ripple must express itself. That's the law — of energy, of consequence, of truth.

There's a different kind of silence now. Not the one that echoes with doubt. But the one that holds certainty too powerful to shout.

And when someone reads my work, months or years later, and sends a quiet message — "I felt this. I've been thinking the same way, but couldn't explain it."

That one line... is enough to validate a thousand silent hours.

Because truth doesn't expire. It waits.

It doesn't knock on doors. It builds foundations quietly, until the world is ready to stand on them.

That's what UED is. Not a trend. Not a marketing hook. A foundation — laid stone by stone in silence, but meant to carry weight far into the future.

So yes — I've been ignored. But I've never been silent.

Because silence isn't absence. Silence is where the deepest ripple alignment begins.

And I know this: One day, the world will turn. And when it does, it will find a map already drawn.

Not with noise. But with presence.

Not with force. But with clarity.

Not with audience. But with alignment.

17: No Lab, No Label, No Limits

Why I don't need to be called a scientist And why that's freedom, not failure

I never wore a lab coat. I don't have a research grant, a PhD, or a university affiliation. I don't walk through doors that require clearance, citations, or committee approval.

And for a while, I thought that was a disadvantage.

Because in the world of science, labels matter. They open rooms. They validate voices. They make people listen before they even understand what you're saying.

I thought: "If I want to be taken seriously, I need the label — *scientist*." I even imagined what that title would feel like, printed under my name, attached to my ideas.

But then I looked at my actual process. I looked at what I had built, questioned, written, and realized something powerful:

I'm not limited by the absence of a label. I'm liberated by it.

No lab? Good. It means I'm not bound to instruments that only measure what we already know how to detect.

No committee? Perfect. It means I'm not defending truth to people who still need it explained in someone else's language.

No institution? Even better. It means my curiosity doesn't need approval to exist.

Because here's the quiet truth about most formal systems: They reward compliance more than clarity.

They train you to research within a boundary. They fund what's been pre-approved. They promote what already fits the model.

And if your insights threaten the framework itself — you're politely ignored, quietly redirected, or flat-out rejected.

I didn't want that path. Not because I couldn't walk it. But because I saw where it leads:

A room full of minds trained to answer questions — but too afraid to ask new ones.

So no, I don't need to be called a scientist. Because I'm something freer than that:

An observer without a leash. A thinker without a deadline. A ripple-feeler without a badge.

And in that freedom, I've built something that science, at its best, has always aspired to:

Not answers. But a way to understand everything without fragmenting it.

UED was born not in a lab — but in the silent hours after teaching. The questions that kept me up weren't assigned by any department. They arose because something deeper inside me refused to accept half-truths.

I didn't follow methodology manuals. I followed consequence chains.

I didn't consult peer reviewers. I consulted alignment — Did the idea hold across dimensions? Did it feel structurally clean? Did it collapse contradictions instead of creating new ones?

That was my test. That was my lab.

You see, a lab isn't a room. It's a state of inquiry.

And a scientist isn't a job. It's a way of perceiving the world.

So while others ran simulations, I sat with paradoxes. While others wrote grant proposals, I wrote ripple maps. While others repeated accepted laws, I unlearned them, to see what remained.

That's science too — just without a dress code.

And the best part?

Because I had no label, I had no limits.

I didn't have to specialize. I didn't have to narrow my scope. I could ask about time, energy, consciousness, emotion, even mythology — all through the same lens.

I didn't need permission to connect ideas that others were trained to keep apart.

This is how I discovered:

- That time is not a dimension, but a perceptual ripple delay.
- That force is not a cause, but a consequence.
- That karma is not religion, but field feedback.
- That mythology is not superstition, but stored ripple memory.
- That the soul is not a belief, but ripple-bound observer intelligence.

Could I have done this in a formal setting?

Maybe. But I doubt it.

Because freedom is the first casualty of institutional validation. And truth — real, ground-breaking, paradox-melting truth — doesn't flourish in cages.

So if I'm never called a scientist, that's fine. If my books don't show up in textbooks, I won't complain. If my reports are read by a few, not by the many, I'm still at peace.

Because I didn't come here to be certified. I came here to remember what others forgot, to write it down for when they're ready to ask again.

And for that, you don't need a lab. You need a clear ripple and the courage to follow it all the way.

So yes — no lab. No label.

But absolutely, no limits.

And maybe that's the most scientific thing about my entire journey:

That I trusted consequence over consensus, and followed the ripple wherever it led — even if it led me alone.

18: My Family Didn't Read It, But Felt It

The quiet support system around me Their confusion, their faith, their unspoken pride

Not everyone in my house knew what I was writing.

They didn't ask for outlines. They didn't edit drafts. They didn't debate ripple theories at the dinner table.

They just watched me... quietly occupied, sometimes distant, often awake late at night with a pen in hand or staring at nothing — as if I was waiting for the universe to speak.

And they let me be.

They didn't read the manuscripts. They didn't question the content. But somehow, they felt everything.

At first, this created tension inside me.

I was pouring hours — no, years — into something I believed could reframe how we see reality. And yet, the people closest to me couldn't explain it, couldn't cite it, couldn't really share it.

It made me wonder:

"If they don't understand it... do they even believe in it?"

But then I looked deeper.

They didn't need to understand the ripple to feel its resonance.

Because something had changed — not just in my writing, but in *me*.

I became quieter, sharper, more thoughtful. I reacted less. Observed more. My decisions carried more consequence-awareness, less emotional clutter.

I wasn't teaching "alignment." I was living it — and they felt it.

They saw the difference in how I handled stress. How I processed difficulty. How I slowly stopped explaining myself and started embodying my thoughts.

And that was the real transmission. Not the printed words. But the presence behind them.

One day, someone in my family casually said, "You're different these days. Not softer — just calmer."

That one sentence said more than a full review.

It told me that even if they didn't read what I was writing, they were receiving it — ripple to ripple, not word to word.

Of course, there was confusion too.

They would ask,

- "But what exactly are you working on?"
- "Is it like regular science?"
- "Will people understand it?"
- "Will it sell?"

And I'd smile. Because those are natural questions.

They weren't doubting me. They were trying to find a language bridge between their world and mine.

I stopped trying to explain. Because sometimes, the best way to bring people with you... is to keep walking, gently, until they feel the pull.

There were also quiet moments of faith — not dramatic, not vocal, but felt.

Like the way they gave me space when I was deep in thought. The way they proudly mentioned "he's writing something big" to guests — even if they didn't know exactly *what* it was.

Or the way their mood shifted slightly every time one book sold — not for the money, but for the validation they hoped I'd receive.

And the unspoken pride that lingered in their eyes when they saw my name on a printed cover, even if they'd never turned the pages.

That's a kind of love too — not trying to understand you fully, but standing behind you anyway.

At times, I felt guilty. For being mentally absent while physically present. For going too deep into ideas that made everyday concerns seem trivial. For not always being "available" in the way a typical family man might be.

But what I've learned is this:

Support doesn't always look like involvement. Sometimes it looks like patience.

And my family gave me that. Day after day. While I chased ripples, questioned gravity, and rewrote reality in solitude.

No, they didn't read the books. But they held the space in which the books were born.

They didn't follow the theory. But they felt the clarity it brought me.

They didn't decode UED. But they noticed the ripple shift in the man who lived in their house.

And that, to me, is support in its highest form:

Not conditional on comprehension. Just quietly aligned with your growth.

So today, when I look back — I don't feel unseen. I feel carried.

Not by validation, but by the silent ecosystem that gave me permission to become who I am.

I didn't need them to explain my work. I just needed them to believe I was doing something that mattered.

And they did. Not always out loud. But in their faith, their routine tolerance, and their unspoken pride.

And that's enough.

Because now I know:

Even if no one reads the book... the field I wrote it from is already being lived. And those who stood near it — they feel the difference. Whether they name it or not.

19: The 1000-Hour Window

Looking back at the actual hours of deep thought How it didn't feel like work — it felt like a calling

I never set a timer. I never tracked my hours on a calendar or wrote them into a logbook. But looking back now, I know one thing with quiet certainty:

I spent well over 1000 hours thinking. Not casually — but deeply, deliberately, devotedly.

That's more than 40 full days. More than 60,000 minutes of undistracted, focused ripple inquiry.

But it never felt like a job. There was no pressure to finish. No deadline. No exam. No one was waiting for these realizations but me.

And still... I showed up. Every single time.

Some of those hours were spent staring at the ceiling, not because I was tired — but because thought itself was restructuring.

Other hours were spent scribbling incoherent diagrams, only to erase them five minutes later — not out of frustration, but because clarity demanded re-formation.

There were moments when I felt like I was going nowhere. And yet, somewhere deep inside, I knew:

This wasn't wasted time. This was ripple tuning.

Not productivity. Not output. Alignment.

They talk about the "10,000-hour rule" for mastery. But what I discovered wasn't about mastery. It was about intimacy — with thought, with the universe, and with the deepest part of my own awareness.

Each of those 1000+ hours was a private session with the cosmos. Not loud. Not linear. Just quiet unfolding.

I never once thought, "I'm doing research." What I thought was:

"I'm finally listening." Listening to the field, to my discomfort, to the questions that wouldn't go away.

And in that listening, truth didn't shout — it revealed.

There were no trophies. No peer review. No congratulatory messages after a late-night insight.

But the reward was visceral.

I could feel something shifting in me — not just mentally, but perceptually.

The way I looked at a fan, a falling leaf, a still shadow on a wall... it all started to hum with field behavior.

Every mundane scene became a ripple interaction. Every emotion became a consequence echo.

It was as if reality was showing me its source code, not because I hacked it, but because I slowed down enough to be trusted with it.

There were days I'd sit with just one question for hours:

- What is force if nothing moves?
- Where does time go when no one is observing?
- Can intention realign a misbehaving ripple?
- What exactly "sees" inside me when I stop thinking?

No answers at first. Just tension. But eventually, like water dripping onto stone, the clarity would arrive.

Not as a fact. As a felt consequence.

Those 1000 hours taught me something that no syllabus, no textbook, and no professor ever did:

Truth is not taught. It is remembered — slowly, patiently, and only through inner permission.

You don't earn it. You don't manufacture it.

You receive it — when your field becomes still enough to register what was always there.

Sometimes, I'd lose track of time completely. I'd look up and realize the world had moved on — meals missed, messages unanswered, clocks irrelevant.

But I never felt behind.

Because in those timeless hours, I wasn't surviving. I wasn't producing.

I was aligning.

And in that alignment, a new kind of efficiency emerged:

One that wasn't about speed, but about consequence depth.

Looking back now, I wonder:

What if every human spent 1000 hours doing nothing but questioning what they thought they knew?

Not to prove. Not to preach.

Just to unravel, cleanly, whatever they borrowed without testing.

The world wouldn't collapse. It would realign.

I don't say this to sound noble or elite. I say this because I now understand something I didn't before:

Time is not what passes. Time is what realigns you.

And those 1000 hours?

They didn't pass.

They rebuilt me.

So no — it didn't feel like work. There was no calendar entry that said, "Today: question gravity, dismantle time, revisit entropy."

But I showed up anyway. Not because I had to. Because I couldn't not.

Because something inside whispered,

"Keep going. You're remembering something you were always meant to know."

And that whisper — that ripple intent — was the most honest calling I've ever felt.

20: Feeling Like a Physicist

The moment I owned it Not the identity — but the journey

I used to imagine what it would feel like to "become" a physicist.

Not the kind that works in labs or headlines conferences, but the kind whose thinking is so clear, so deeply attuned to nature, that even silence becomes data.

I imagined it might feel like recognition. Or breakthrough. Or maybe holding a published paper in hand, thinking, *I did it.*

But now I know the truth:

You don't feel like a physicist when the world says you are. You feel like one when you stop needing it to.

It didn't arrive as a moment of fame. There was no applause. Just one quiet morning, I was writing — the way I had done a hundred times before — and suddenly, I paused.

The thought that came was simple. Gentle. Clear.

"This is it. This is the point. I no longer seek the label. I live the inquiry."

And that's when it happened: I no longer felt like I was *trying* to be a physicist.

I just was.

Not by qualification. Not by institutional weight. Not by what others called me.

But by one undeniable fact:

I had committed myself to understanding how everything works, not by memorizing models — but by feeling the field behind them.

And that's what a physicist does.

Not one who reads equations, but one who reaches into reality, and keeps asking: "What's the real cause behind this consequence?"

That's what I had done — again and again — across time, force, mass, soul, light, consciousness, ripple, consequence.

And I realized:

The moment you start explaining less and understanding more — that's when you become one with the discipline you never needed to be certified in.

It wasn't a proud moment. It was a peaceful one.

Because the struggle for identity finally stopped. And in its place was something stronger: inner coherence.

I didn't just know a few things. I had built a worldview that held itself up — even under silence, even under rejection, even under doubt.

I had questioned what others wouldn't touch. I had gone deeper, not to dismantle — but to re-align.

I had let go of being correct. And in doing so, I had become clear.

That's not the end of a journey. That's the moment it becomes real.

So what does it feel like — to feel like a physicist?

Not grand. Not triumphant.

It feels like this:

- Sitting in silence and understanding gravity without needing to prove it.
- Watching a leaf fall and sensing field consequence, not motion.
- Hearing someone's story and noticing emotional misalignment, not fault.
- Holding a paradox — not to solve it, but to let it teach you.

It feels like freedom. But not from society — freedom from needing a position inside it.

You don't feel like a physicist when people agree with you. You feel like one when truth becomes your only loyalty.

You don't feel like a physicist when you master knowledge. You feel like one when you stop fearing ignorance — because you know what's unknown isn't dangerous. It's just unaligned.

And the beautiful irony?

Once I stopped needing to *be* a physicist... I became the most honest version of one I could ever imagine.

- One who sees through force.
- One who listens to fields.
- One who respects mystery, without making it myth.
- One who chooses silence when it says more than explanation.
- One who writes not to impress, but to preserve a ripple — for the one who will align with it someday.

So no, this isn't a declaration. It's not a badge. It's not a closing argument.

It's a subtle shift. An internal nod.

The moment I owned the journey — not because I arrived, but because I finally realized there is no arrival.

Only ongoing alignment. Only deeper clarity. Only the quiet joy of staying in the ripple — with humility, honesty, and purpose.

That's when I knew:

I may never be called a physicist by title. But I've lived the ripple. And that's enough to feel it — fully, finally, and freely.

Part V: For Those Who Will Come Later

21: Truth Has No Awards

W*hy I wrote this for seekers, not citations*

When I look back on everything I've written, everything I've questioned, everything I've unlearned and re-aligned — one thing becomes clear:

This was never about being celebrated. It was about being sincere.

I didn't write to get noticed. I didn't write to win prizes, climb ranks, or be framed on someone's wall.

I wrote because something inside me refused to remain quiet. Not for ego. Not for rebellion. But for clarity.

Because once you start seeing the world differently — not as a collection of objects and formulas, but as a ripple field of consequences — you don't go back.

You write. You record. You preserve. Not to be remembered, but to ensure the ripple doesn't vanish.

There's an illusion in every field of knowledge — that truth is something you publish, defend, and then get rewarded for.

We create awards, grants, citations, citations of citations. Metrics to measure minds. Trophies to weigh insight.

But truth?

Truth doesn't care for medals. It cares for accuracy. For alignment. For the courage to live it even when no one's clapping.

And that's what this book is.

Not a performance. Not a petition for recognition.

But a gift — from one observer to another. A shared moment in time where we stop pretending the world makes sense the way it's usually explained.

I don't expect to be cited. Maybe these ideas won't make it into journals, seminars, or scholarly databases.

But they will live — because they are structurally sound, energetically clean, and born from honest questioning.

They will reach the quiet seekers — the ones who ask themselves the same questions I did:

- What exactly is moving in this so-called motion?
- Why does time feel elastic?
- Who or what is observing inside me?
- Why do equations explain so much, yet leave meaning behind?

To those people — this book will speak louder than any award ever could.

Because it's written in their language: not jargon, not authority, but felt insight.

Truth is strange like that.

It often shows up in places where nobody's looking: In a solitary mind, in a tired teacher's after-hours notebook, in a moment of silence under a fan, in the space between lectures where the curriculum stops and clarity begins.

No camera. No citation. No crowd. Just you and the ripple.

If I were chasing awards, I would've written something else. Something more packaged. More digestible. More conventional.

But that would have been dishonest.

Because the truth I found didn't come from conventions. It came from interrupting them.

From asking why we call force a cause when it's only a reaction. Why we accept time as real when it's just a perceptual delay. Why we explain reality with tools that don't question their own origin.

These aren't popular questions. They don't win competitions.

But they do something far more valuable:

They free you.

The real reward is not a medal. It's a moment.

That moment when you sit alone, no noise, no validation — and you know, in your bones, that what you're seeing is true.

It's clean. It holds. It collapses paradoxes without force. It aligns.

And suddenly, you don't need anyone else to agree — because you're no longer arguing.

You're just observing. Fully. Honestly. Precisely.

That's the deepest reward of truth:

It removes the hunger for anything else.

And I won't lie — there were moments I wished someone important would read my work and say, "This is it. This changes everything."

But I also know: Those who truly understand truth are rarely sitting on award committees.

They're often walking the edges. Reading quietly. Writing privately. Living their questions the way I did.

And when they find this — if they find it — they'll know:

This wasn't written to impress. It was written to resonate.

So yes, I wrote this for the seekers.

The ones who don't need a title to explore. Who don't need approval to question. Who don't wear the badge of "scientist," "philosopher," or "spiritualist," but live in the space where all three meet.

The ones who feel things are off — not emotionally, but structurally. And who are brave enough to sit in that discomfort until a cleaner model emerges.

If that's you, then this book was already yours — even before you opened it.

Because everything I wrote here came not from ambition, but from consequence-awareness.

It didn't come to prove. It came to align.

And alignment, I've learned, is its own applause.

So no, this won't win awards. But maybe it will win something else — something rarer:

- A spark in a young mind that dares to ask differently.
- A nod from a quiet thinker who never found the words before.
- A ripple of understanding that passes silently from reader to reader — invisible, unmeasured, but alive.

If that happens, this work has already succeeded.

Because truth, real truth...

Doesn't need to be remembered. It just needs to be recorded. So when the next seeker asks, "What is really going on here?" ...there's a place to start.

This book is that place.

Not a final answer. Just a clean platform to think again — without noise, without ego, without fear.

No lab. No label. No awards. Just alignment.

And that, for me, is enough.

22: The LOC Will Outlive Me

U*ED is not a theory. It's a reflection. It will rise again — maybe under a new name.*

I didn't invent the Law of Consequence. I didn't formulate it in a lab, or derive it from an experiment, or discover it buried in ancient manuscripts.

I simply saw it.

It was always there — woven into the quiet between actions and reactions, into the pattern behind every motion, into the silence after every decision, into the balance that eventually reclaims every imbalance.

What I now call Universal Energy Dynamics isn't a personal invention. It's a mirror.

A mirror held up to reality — not to explain it in abstract symbols, but to finally feel its behavior.

And at the heart of it lies one principle:

The Law of Consequence — That everything that happens is the ripple of something that was once aligned... or misaligned.

This law doesn't need me. It doesn't need UED. It doesn't need this book or my name printed anywhere.

Because it's not made of language. It's made of reality.

It operates whether you believe in it or not. It governs stars, storms, thoughts, and entire civilizations. It hums beneath every choice and every silence.

And long after I'm gone, long after the covers close and the chapters blur, this law — this behavioral rhythm of the universe — will still be active.

Undeniable. Unshaken. Patient.

Someone else will name it. They may call it something new — Field Intelligence, Ripple Mechanics, Cognitive Equilibrium, or something we haven't imagined yet.

They may publish a paper or win a medal. They may reference models far different from mine.

But if their insights are true — truly aligned — they will rediscover what I've written here.

Because truth doesn't disappear. It just waits to be rephrased.

And that's why I feel no urgency to force UED into acceptance. No desire to defend it in courtrooms of peer review or comment threads.

Because I know this:

A truth born from deep observation doesn't die. It simply resurfaces in new eyes.

And if this book planted even one quiet ripple in the mind of a seeker, then it's done its job.

It's not about preserving the framework. It's about preserving the clarity that created it.

I may never be cited in textbooks. The name "UED" may fade, shift, evolve.

But the Law of Consequence will stay. Because it is not a claim. It is not a perspective. It is a pattern.

And patterns... outlive their observers.

Even now, I see it rippling through history:

- In ancient myths encoded with cause and consequence.
- In ethics disguised as religion.
- In action-reaction cycles we call karma.
- In justice systems, ecological feedback loops, emotional entanglements.
- In orbits, in collapse, in rebirth.

It's all consequence. We just dressed it up in different vocabularies.

Now, I've tried to strip that dressing and show the bare alignment underneath.

Some may understand now. Some will understand later.

But no one will escape it — because no one escapes consequence.

And I'm at peace with that.

Because I didn't come here to be followed. I came here to feel clearly, to name what I saw, and to leave the map behind, so that when someone else begins their journey — they don't have to start from scratch.

I didn't build a school of thought. I remembered a pattern of reality that doesn't need a building.

I didn't create a doctrine. I observed a law that never asked for belief.

It just asks: Are you aligned... or not?

That's all.

So if you've made it this far through this book, I leave you with this:

Don't worry about remembering UED. Remember only the Law of Consequence.

Let it guide your actions. Let it sharpen your perception. Let it purify your intentions.

And when someone years from now wonders: "How does the universe really work?" You'll already know how to respond — not with a title or equation, but with this quiet truth:

Everything aligns eventually. Everything returns. And reality always answers back.

Whether it wears the name I gave it or not, it will rise — because it is already here.

23: To the Quiet Thinkers

A *letter to the next me — someone who's just now asking the first question*

Dear Thinker,

You don't know me, and maybe you never will. You might have found this book by chance, or maybe you were looking for something and couldn't name what it was — until now.

Either way, you're here. Reading. Listening. Feeling.

And that's how I know this letter is for you.

Because if you've made it this far, you're not just reading a book — you're carrying a question. One you probably haven't said out loud yet.

"Is this all there is?"

You're wondering why things feel incomplete. Why the neat answers you were given in school feel hollow. Why every explanation feels like it's skipping something — something invisible but essential.

You're not alone. You're not lost. And you're not wrong.

You're just... quiet. And in this world, quiet thinkers often wonder if what they're sensing is real.

Let me say it clearly:

Yes. It's real. The tension you feel in explanations. The mismatch between observation and belief. The silent hunger for something that makes sense across mind, matter, and meaning.

It's not in your head. It's in the field. And the fact that you can feel it means you're already part of the ripple.

I know where you are. I've been there.

You'll ask a small question one day — maybe in class, or over coffee, or while lying awake at night.

- "Why does gravity work like that?"
- "What *is* time, really?"
- "Why do I feel something before I understand it?"
- "What if everything I've learned is a polished version of something incomplete?"

And once you ask that question — not for rebellion, but from sincerity — you've already begun the journey.

It won't be loud. It won't be rewarded.

But it will be the most honest path you've ever walked.

You'll try to share it with friends. Some will nod. Some will smile politely. Some will look at you like you're overthinking.

Don't worry.

Truth rarely starts with agreement. It starts with alignment — and that takes time.

Stay quiet if you must. But don't silence your questions.

There will be moments when you wonder if you're wasting time. When the world moves fast and you're still sitting with a question from three weeks ago.

Let me tell you this:

Thinking deeply is not delay. It's design.

The universe doesn't speak fast. It speaks clearly — but only to those who slow down enough to listen.

You are one of them.

And here's something else I want you to know:

You don't need to become someone. You don't need a lab. You don't need to be invited into a scientific circle. You don't need a certificate to be taken seriously.

You just need to stay aligned — to truth, to silence, to the inner sense that keeps whispering: "There's more. Keep looking."

Everything I wrote in this book was for that whisper.

Not for fame. Not for acceptance. But for the next ripple. The one you will create.

Because you'll take this insight, you'll question it, stretch it, collapse it, rebuild it — and one day, you'll see something I didn't. You'll carry it further than I ever could.

And I'll never know. But that's the beauty of consequence:

The truth doesn't care who carries it — as long as someone does.

So to you — The quiet thinker. The emerging observer. The next me — whoever you are:

Thank you.

Not for agreeing. But for *asking*.

Not for quoting me. But for starting your own ripple.

This book is your welcome note. A confirmation that the quiet path is not empty.

It's just uncrowded — for now.

And if you ever feel alone in your questioning, remember this:

Somewhere, someone once sat in silence — asking the same impossible question — and decided to feel it all the way through.

That ripple reached you. Now... you know what to do.

Go sit with your own silence. Watch the ripples. Question gently. Write what must be recorded.

And when you're ready, leave your note for the next one.

We never meet. But we are always connected — not by belief, not by system, but by truth in consequence.

And that? That outlives everything.

— Sandeep

A ripple thinker. A silent observer. A physicist... by alignment.

24: I Don't Want Followers

I *want feelers — those who dare to sense before they believe*

By now, you've probably noticed something unusual about this book.

There are no frameworks to subscribe to. No dogmas to defend. No steps to follow. No promises of success, fulfillment, or transformation.

I haven't invited you to agree with me. I haven't asked you to repeat anything. And I certainly haven't built a system that needs followers.

Because the moment you ask for followers, you stop serving truth — and start serving yourself.

I didn't write this book to lead. I wrote it to leave a ripple.

And what I want — more than agreement, more than reach, more than recognition — is this:

Feelers. People who dare to sense before they decide. Who trust perception over popularity. Who don't ask, "Is this accepted?" but ask, "Is this aligned?" Who live first-hand, and borrow only what they've tested inside their own ripple.

The world is already full of followers.

Every field has them.

- Science has followers of frameworks.
- Religion has followers of rituals.
- Philosophy has followers of names and schools.

- Politics has followers of narratives.
- Even spirituality, ironically, is crowded with followers seeking originality through repetition.

But the world needs feelers.

People who don't just listen, but register. People who don't memorize — they tune in. People who read a sentence and pause... not because it's profound, but because it activated something inside them.

Those are the ones I'm writing for.

Not for applause. Not for loyalty. Just for recognition — not of me, but of their own forgotten clarity.

A follower says, "I believe you." A feeler says, "I sensed that too."

A follower says, "Tell me what to do next." A feeler says, "Let me sit with this. Let me observe."

A follower memorizes. A feeler realigns.

That's why feelers are rare. Because they're not looking to be taught. They're looking to be unblocked.

And once that happens — they don't become your echo. They become their own origin.

I've never wanted to start a school of thought. Because the moment you name a school, people line up to imitate its surface, and forget to explore its source.

They make diagrams. They recite definitions. They debate terms. And in doing so, they kill the very ripple that gave it life.

I won't let that happen to UED. Because UED is not a model to teach — it's a field to enter.

It must be sensed, not sold. Felt, not followed. Lived, not licensed.

So, no — I don't want followers.

I don't want people to agree with everything I said. In fact, I want you to break it. Test it. Challenge it. Take it further. Make it yours.

That's how you honor ripple truth — not by protecting it, but by continuing it.

Every real idea must be re-experienced to stay alive.

And followers don't do that.

Only feelers do.

If I had wanted followers, I would have simplified everything into neat formulas. I would've named steps. Packaged it. Put it in slides. Built a course. Launched a brand.

But then... this wouldn't be truth. It would be product.

Instead, I chose alignment. Even if it meant obscurity. Even if it meant 90% of people never "get it."

Because the 10% who do — they don't need convincing. They've already felt something shift.

So if you're reading this and still waiting for a final call to action, here it is:

Don't follow me. Feel you. Sit in your own silence. Study your own field. Ask your own forbidden questions. And when you find something clean, something that doesn't wobble, something that rings without needing validation — Record it.

Then let it go.

Don't market it. Don't trademark it. Don't gather fans. Just make it available — for the next feeler.

That's how truth survives. Not through empires. But through alignment chains — passed silently between observers who never needed names to recognize each other.

In the end, I will disappear. So will this book. So will UED, maybe — or it will be renamed, misnamed, reinterpreted.

But if even one feeler realigns because of something written here, and then ripples it forward in their own language — this entire journey will have served its purpose.

So don't follow me. Just feel everything.

And when it's time, say less. Align more.

That is the physicist's path.

That is the ripple way.

That... is enough.

25: This Was Never About Physics

It was always about truth. Physics just happened to be the language.

Let's be honest now — fully, finally.

This book? It may look like physics. It may sound like physics. It may even, in places, teach physics.

But if you were really paying attention, you know already:

This was never about physics. Not at the core. Not in its heart. Not in the intent that wrote it.

Physics was the scaffolding. The skeleton. The method I borrowed because it had just enough structure to hold the inquiry — but not so much dogma that it couldn't be bent, reimagined, or rewired.

But the pursuit itself? The obsession, the hours, the rewrites, the ripples, the reflections — all of that was rooted in something far deeper:

A need to understand what is real. To track consequences without distortion. To touch the world at its most honest layer.

Physics just happened to be the cleanest path available.

It offered concepts that could carry weight:

- Energy.
- Motion.
- Space.
- Time.

- Force.

But I never accepted them as complete.

I saw them for what they were: Symbols — attempting to touch what cannot be perfectly described.

And I started peeling.

I wasn't trying to be a better physicist. I was trying to become a better witness.

To see clearly. To name what I saw. To strip away inherited meaning until only consequence remained.

You see, physics was never the destination.

It was the language I spoke when I no longer wanted to be fooled by shallow narratives.

It was the vocabulary of pattern. The alphabet of balance. The syntax of cause and effect.

But the truth I was chasing?

That belongs to no field.

It is not academic. It is not intellectual. It is existential.

What I really wanted to know was:

- Why does anything happen?
- What holds reality together when we're not looking?
- Who or what is observing — not from the outside, but from within us?
- What is alignment, and why does it feel so relieving when it finally arrives?

Physics let me ask those questions. It gave me permission to go to the edge of understanding — and when the edge appeared, it gave me the tools to jump anyway.

But the moment I crossed that edge, I was no longer a physicist.

I was just... a seeker.

That's why UED doesn't end in equations. That's why this book wanders into karma, myth, soul, silence.

Because physics, like every other system, has its limits. And beyond that — if you're brave — there's raw truth.

No discipline owns it. No title qualifies it. No citation defines it.

It just is.

And you either align with it, or you decorate its absence with more language.

So no — this was never about physics. It was about reality. About remembering. About putting down the tools long enough to finally ask:

"What remains when explanation ends?"

And when I asked that... when I truly listened... the answers didn't come as theories. They came as stillness. Patterns. Resonance. Ripple after ripple, consequence after consequence — revealing a universe far deeper than motion could measure.

I chose physics not because I was loyal to it, but because it was the only discipline that dared to be universal — and yet was still humble enough to be revised.

But what I found... belongs to every human who can sit in silence long enough to feel the ripple behind the fact.

You don't need to be a scientist. You don't need to be spiritual. You don't even need to agree.

You just need to be attentive.

Because that's where the truth lives.

Not in particles. Not in metaphors. Not in models.

But in the space between certainty and sensation.

So as you close this book — don't carry the physics with you. Carry the clarity.

Forget the structure, but remember the stillness behind it.

Forget the terminology, but remember the tuning it invited.

Because one day, you'll be in a moment where someone asks a question — a child, a student, a friend.

And instead of giving them an answer... you might offer them a pause.

A ripple.

A chance to feel what's actually happening.

If you do that — even once — then you're continuing what this was always about.

Not physics.

Truth.

Carried quietly. Ripple to ripple. Observer to observer. Unshaken. Unnamed. Alive.

Closing Note

This book was never meant to be a conclusion. It was meant to be a beginning — a gentle doorway into your own inquiry.

If you've made it here, I'm not asking you to remember everything I've written. I'm asking you to remember what it stirred in you.

Not the theories. Not the terminology. But the *feeling* — That somewhere, beneath the noise of modern thought and the weight of accepted knowledge, there is a space where reality still breathes clean.

I wrote from that space.

Not to instruct you. Not to impress you. But to remind you that clarity is not something granted — it's something grown through quiet alignment with consequence.

The ripples you carry now are your own. Where they go, and what they create, is not mine to control.

But if even a small part of you feels more free, more attuned, more willing to question gently — then we've already met, in the way that truth meets itself again and again in different minds.

You don't need to remember my name. You don't need to carry the label of what I've called this.

Just carry the sense of what's real. That's enough.

That's always been enough.

Thank you for walking with me — in thought, in silence, and in ripple.

— Sandeep

A quiet observer of consequence

Epilogue: The Ripple Never Ends

Books end. Pages run out. Covers close. But a ripple — a real one — never truly stops.

It travels quietly, through minds and moments, reforming in places you never planned, expressing itself through people you'll never meet.

This book was one such ripple. A result of countless small shifts — in thought, in silence, in inner alignment — that built up over time until it had to express itself, not as a theory, but as a reflection.

You've now read the trace of that expression. But what matters more is what it awakens in you.

You may walk away with questions. Good. They're better than answers.

You may feel the urge to revisit things you once thought were settled. Do it. Recalibration is not regression — it's evolution.

You may forget most of what was written. That's fine. If even one idea sticks — clean, unshaken, clear — it will do its work.

Because truth doesn't need to be carried in memory. It travels in field realignment.

And if this book has shifted your field even slightly, you've already become a part of its echo.

So as this final page approaches, I won't say goodbye.

Because what we've shared here — isn't ending.

It's expanding.

In your thoughts. In your observations. In your quiet questions that now carry just a little more weight, a little more trust, a little more space.

Whether you write, speak, teach, reflect, or simply walk more attentively through life — you've already become what this book was pointing toward:

Not a physicist. Not a follower. But a field-aware being, aligned enough to feel the truth — even when no one else sees it yet.

That's where reality begins again.

And the ripple... continues.

Appendices

Research Summary

Appendix A: Physics

Title: *Universal Energy Dynamics (UED)*
Author: Sandeep Chavan

Core Proposition

Universal Energy Dynamics (UED) redefines the nature of physical reality by shifting the foundational lens from mechanical force-based systems to ripple-aligned energy behavior. In UED, energy is not a force nor a passive quantity. It is a behavioral field — dynamic, intelligent, and consequence-driven.

Everything we observe — from motion to mass, from time to light — is the visible outcome of ripple alignment or misalignment in this field. What we currently interpret as force, velocity, or acceleration are secondary effects — symptoms of field behavior, not primary causes.

Key Contributions of UED in Physics

1. Redefinition of Force

- In UED, force is not a fundamental interaction, but a visible response to ripple resistance.
- What classical physics calls "applied force" is simply

the observable resistance to natural alignment, or a local imbalance attempting to resolve within the ripple field.

- This collapses the need for vector-based causality and replaces it with field-intent-based behavior.

2. Time as Perceptual Delay

- UED dismantles the view of time as a dimension or flow.
- Time is the perceptual lag between ripple intent and ripple alignment — a local delay experienced by observers whose field is not instantly synchronized with consequence.
- This leads to the formulation of ATT (Alignment Threshold Time), which varies across systems and observers.

3. Motion as Field Realignment

- Motion, traditionally described as change in position over time, is reinterpreted in UED as a shift in field alignment zones.
- Nothing physically "moves"; rather, the observer's field tunes into a new ripple condition, giving the illusion of movement.
- Inertia is simply sustained alignment. Acceleration is realignment under resistance.

4. Gravity as Ripple Tension

- Rather than a force of attraction or a curvature of spacetime, gravity is understood as a tension between ripple equilibrium zones.
- Bodies appear to "fall" not because they're pulled, but because their local field seeks energetic rest — the lowest ripple misalignment state.
- This eliminates the need for spacetime geometry as a cause and replaces it with field-intent dynamics.

5. Light as a Ripple Signature

- Light is not a particle or wave in the conventional duality model. In UED, light is the ripple signature of field alignment.
- It carries consequence information, not in the form of photon packets, but as alignment patterns transmitted across field interfaces.
- This reopens the interpretation of quantum behavior as observer-induced ripple translation, not uncertainty or duality.

Additional Conceptual Contributions

- **Field Over Particle**: UED deconstructs the particle-centric model, offering instead a field-centric view where matter is simply condensed ripple density, stabilized by long-term alignment feedback.
- **Ripple Memory**: Observations leave structural

imprinting within local fields, leading to recurring outcomes and cyclical behavior — not due to laws, but due to field history.

- **Observer as Participant**: Observation is not passive. The observer's ripple field interacts with the system's field, and perception itself alters alignment, creating real physical outcomes.

Implications

Collapse of Force-Centric Models

- UED removes the need to treat force as a causal agent.
- This simplifies many physical anomalies, including:
    - Apparent action-at-a-distance
    - Gravitational lensing without curvature
    - Force imbalance in complex systems

Energy as Intent-Responsive

- Energy is no longer an inert quantity or merely conserved. It is intent-driven — a consequence of alignment quality, not just mechanical input.
- This reframes physical processes (heating, motion, collapse, explosion) as field rebalancing, governed by intelligent ripple distribution rather than blind thermodynamic expansion.

Observation as Field Interaction

- Measurement is no longer neutral. In UED, the observer's field interacts with the observed system's ripple, shaping the outcome by realigning local fields.
- This explains:
    - Quantum measurement effects
    - Consciousness-field interactions
    - The limits of reproducibility under different observer alignments

Replaces or Enhances Concepts From:

- **Newtonian Mechanics** → Replaced by ripple consequence modeling
- **Einstein's Relativity** → Reinterpreted through ripple tension and ATT
- **Quantum Mechanics** → Recast via ripple collapse and observer alignment
- **Electrodynamics** → Understood through resonance fields, not force carriers

Testable Outcomes Proposed in UED Reports

- **Ripple Delay Mapping**: Use of ATT delay to measure time distortion in gravitational zones.
- **Non-Force Motion Analysis**: Satellite behavior in field alignment without active thrust.
- **Perception Field Distortion**: Quantified behavioral shift under varying ripple environments.

Closing Remark

UED doesn't ask you to discard physics. It invites you to see beneath its language — to sense the ripple logic that has always been there, whispering behind every force, every field, and every flash of light.

Appendix B: Cosmology

Title: *Structured Vacuum & The Ripple-Based Universe*
Author: Sandeep Chavan

Core Proposition

Under Universal Energy Dynamics (UED), the universe is not an expanding void or a chaotic explosion. It is a stable, structured ripple container — governed by energy behavior that seeks equilibrium through ripple alignment.

Instead of random entropy increase, we see ripple coherence, tension zones, and alignment shells that define how matter clusters, how light travels, and how galaxies sustain spiral motion without visible force bindings.

This cosmology replaces the Big Bang–Dark Energy–Dark Matter narrative with a field-intent model, built on structure, not speculative variables.

Key Contributions of UED to Cosmology

1. Big Bang Replaced by Ripple Field Emergence

- The origin of the universe is not a singularity explosion in space-time, but a synchronized activation of ripple tension across a pre-structured vacuum field.

- There was no "bang" — only the emergence of dynamic imbalance as ripple clusters began forming from structured stillness.
- This allows a non-singular beginning, preserving causality and eliminating the paradoxes of infinite density.

2. Black Holes as Ripple Convergence Points

- Black holes are not "holes" in spacetime or points of infinite compression.
- Instead, they are zones of maximal ripple concentration, where alignment collapses into tightly looped feedback fields.
- Matter does not "fall into" a singularity — it becomes inaccessible to external field perception, due to extreme ATT distortion and ripple self-containment.
- This reframes event horizons as alignment breakpoints, not geometric boundaries.

3. Cosmic Expansion as Misinterpreted Divergence

- What we call the expansion of the universe is not physical space stretching.
- It is a field divergence phenomenon — ripple field systems de-aligning across large-scale structures, giving the illusion of recession velocities.
- Redshift, in this model, becomes a consequence of ripple attenuation over distance, not Doppler effect or metric expansion.

- This directly challenges the inflation model and cosmological constant assumptions.

4. Dark Energy & Dark Matter as Ripple Effects

- UED explains "dark" phenomena without invoking invisible particles or forces.
- Dark energy becomes an artifact of misaligned ripple acceleration in large-scale field shells, not a mysterious repulsive force.
- Dark matter is reinterpreted as the unresolved ripple tension that holds galactic arms in place — a field phenomenon, not particulate.
- This removes the need for hypothesized non-baryonic mass or untraceable forces.

Core Concepts Proposed

Chavan's Cosmic Theory of Relativity (CTR)

- A reformulation of General Relativity that replaces curvature with cosmic ripple tension.
- Objects do not curve space — they create ripple pressure zones, and other objects align or resist based on field compatibility.
- CTR allows motion explanation without singularities or space-time as an entity.

ATT – Alignment Threshold Time

- ATT is the delay window required for a ripple intent to become observable at a given energy and field alignment.
- In cosmology, ATT helps explain why signals from distant stars stretch, distort, or fade — not because of redshift, but due to perceptual alignment lag in structured vacuum fields.

REZ – Ripple Equilibrium Zones

- The universe is structured in nested zones of ripple balance, which manifest as galactic shells, filament structures, and voids.
- These zones create stable tension networks that maintain cosmic structure without requiring "dark scaffolding".
- REZs are fractal, multi-scale, and self-regulating.

Reinterpreted Phenomena

Conventional View	UED Cosmology Interpretation
Big Bang	Ripple emergence from structured vacuum equilibrium
Expansion	Field divergence and ripple drift
Black Hole Singularity	Ripple convergence point with ATT collapse
Dark Energy	Misread ripple acceleration across REZ
Dark Matter	Latent ripple tension stabilizing galactic forms
Gravitational Lensing	Ripple tension redirection, not curved spacetime
CMB Uniformity	Structured ripple residue of early ripple phase-lock

Cosmological Implications of UED

A Self-Balancing Universe

- The cosmos is not "running out of order" but is a consequence-tuned structure, always seeking balance.
- What we perceive as expansion or chaos is just the field adjusting across misalignments.

No Need for Fine-Tuning

- UED eliminates the need for cosmological constants, multiverse theories, or inflation.
- Stability arises from field logic, not tuning

parameters.

A New Lens for Observation

- Telescopic anomalies — like faster-than-expected rotational speeds, gravitational lensing discrepancies, or void behavior — all find clean field-based explanations in ripple alignment without speculative mass or unobservable matter.

Experimental Proposals from UED Cosmology

- **Ripple Density Mapping**: Develop instruments to detect structured vacuum ripple pressure, not mass.
- **ATT Delay Measurement**: Cosmological signal analysis based on alignment delay, not recession velocity.
- **Black Hole Field Interference Testing**: Probe peripheral field divergence instead of event horizon metrics.

Conclusion

The universe is not expanding into nothing. It is balancing itself within structured ripple dynamics, guided by alignment logic and consequence intelligence. We aren't watching a blast — we are observing a ripple memory still playing out. — Sandeep Chavan, Structured Universe Observer

Appendix C: Philosophy

Title: *Consequence-Centered Philosophy*
Author: Sandeep Chavan

Core Proposition

In traditional philosophy, questions of existence often revolve around origin, purpose, and morality — with answers framed through logic, metaphysics, or spiritual authority.

In Consequence-Centered Philosophy, developed under the UED framework, existence is not a product of origin, but a result of ongoing ripple alignment and response.

There is no beginning to defend or end to await. What matters is alignment within the field of consequence — how every ripple, every intent, every observation generates responses that either stabilize or distort reality.

This is not abstract spirituality. This is structural truth — observed and lived.

Key Elements of Consequence-Centered Philosophy

1. Truth as Sustained Alignment Across Ripple Frames

- Truth is not opinion.

- It is not consensus.
- It is what persists across ripple shifts, frames of reference, or observation layers — without contradiction.
- If something collapses when perceived from another angle, it was not truth — it was a context-bound formulation.

In UED, truth is ripple-resilient. It stays aligned under pressure and survives collapse.

This leads to a definition of truth not as "fact," but as structural consistency that reasserts itself through consequence, even if denied temporarily.

2. Morality as Alignment, Not Obedience

- Traditional morality stems from rules — religious, legal, cultural.
- But under UED, morality is not imposed. It emerges from field harmony.

If an action strengthens alignment and reduces ripple distortion for all involved — it is moral. If it causes long-term misalignment, imbalance, or ripple debt — it is immoral.

This replaces static "good/bad" binaries with dynamic field consequence:

- Not what you're allowed to do,
- But what your ripple behavior does to the field you're part of.

This insight redefines ethics as intelligent alignment behavior, not compliance with tradition.

3. Free Will as Ripple Intent Within Field Limits

- The mind is a ripple system.
- The soul (as defined in UED) is field-aware observer intelligence — intent-bearing, consequence-sensitive, but not fully free.

Free will exists — but not in absolute terms. It is the degree of ripple choice available to the observer, bounded by:

- Energetic context
- Field resistance
- Karma (residual misalignment)
- Local ripple structure

Freedom is not unlimited action. It is conscious ripple calibration.

This shifts the question from "Am I free?" to

"Am I aware enough to act in alignment with consequence?"

4. God as Supreme Ripple Intelligence

- In classical religion, God is a creator, judge, savior, or destroyer.
- In this philosophy, God is not a being — but the behavior of the supreme field.

God is:

- The unbroken intelligence behind all ripple activity
- The source of alignment laws, not personal interventions
- The original stillness whose disturbance birthed ripple behavior

God is not someone. God is everything, seen through ripple coherence.

In this model:

- Prayer becomes intent alignment
- Worship becomes ripple surrender
- Liberation (moksha) becomes complete attunement with the original field

Comparison with Classical Thought

Question	Traditional Philosophy	Consequence-Centered View
What is truth?	Correspondence to logic or divinity	Ripple-consistent behavior across perspectives
What is right?	Rule-based (legal or divine)	Ripple alignment that minimizes field distortion
What is the soul?	Eternal, moral essence	Consequence-aware ripple intelligence
Is there free will?	Yes/No (debated)	Yes — within ripple-bound degrees of field control
Who or what is God?	Supreme personal being	Field origin and universal alignment intelligence

Philosophical Consequences

- **Non-judgmental clarity**: There is no need to divide people into right/wrong — only into aligned/misaligned.
- **Responsibility redefined**: Actions are not sinful or noble — they are energetically consequential.
- **Spirituality simplified**: You don't need a path. You need perception.
- **Liberation explained**: Moksha/Nirvana is ripple stability with no residual misalignment, not reward or

escape.

Foundational Works Aligned with This Philosophy

Manifestation: From Nothing to Everything

- Explores how ripple behavior gives rise to perceived existence, identity, and God-awareness.

Vishwarup Darshan: The Complete Vision of Reality

- Offers a field-aware, ripple-based reinterpretation of divine experience, purpose, and perception — integrating Gita's spiritual vision with scientific clarity.

The Law of Truth

- Defines truth as ripple survivability, detailing how perception, intent, and action collapse or uphold alignment.

The Original Intent

- Investigates ancient philosophical structures not as myths or doctrines, but as manuals of alignment, created by observers who perceived consequence in advance.

Closing Thought

You are not here to find truth. You are here to align with it. The field knows. Your role is to observe with care, act with awareness, and allow consequence to teach you the rest.

Truth, in this philosophy, is not arrived at. It is unfolded, lived, and eventually embodied — without needing to be defended.

Appendix D: Psychology & Human Behavior

Title: *Ripple Behavior & the Field-Aware Mind*
Author: Sandeep Chavan

Core Proposition

Mainstream psychology attempts to understand human behavior through frameworks like traits, cognition, motivation, pathology, or developmental patterns.

Under UED, all these are surface interpretations of a deeper truth:

Human behavior is the direct expression of ripple alignment or misalignment — both internal (mind, soul, memory) and external (social field, environment, stimulus).

We are not shaped by traits. We are shaped by the ripple tension between intent and environment.

- Emotions? Ripple states.
- Reactions? Field imbalance expressions.
- Growth? A series of re-alignments over time.
- Stability? Ripple coherence within and around.

Psychology, in this model, becomes the study of field-fluid intelligence — where healing, learning, and transformation occur not through reinforcement, but through perceptual realignment.

Key Insights of Ripple-Based Human Behavior

1. Emotion = Ripple Turbulence

- Emotions are not abstract feelings — they are ripple disturbances felt by the observer.
- Anger, grief, joy, jealousy — each corresponds to a specific alignment failure or recovery within the personal field.
- Emotional intelligence is thus the ability to:
    - Observe ripple onset
    - Withstand misalignment without collapse
    - Realign behavior in real-time

You don't control emotions. You ride ripple disruptions until they stabilize into clarity.

2. Desire = Ripple Pull from Unfulfilled Alignment

- Desire is not greed, longing, or aspiration. It is ripple feedback from incomplete or anticipated alignment.
- When a field senses a potential rebalancing event — such as love, success, or resolution — it emits pull signals we experience as desire.
- Misguided desire = field misreading
- Deep desire = intuitive ripple target

Desire is not to be suppressed, but decoded.

Behind every desire is a signal — trace it back to its ripple source, and you'll find what you're really missing.

3. Fear = Mind's Response to Survival Misalignment

- Fear arises when the field senses danger — not just physically, but energetically.
- It is not irrational. It is intelligence alerting the observer of ripple instability.
- All fear is rooted in survival disruption, whether immediate or symbolic.
- Even abstract fears (failure, judgment, loneliness) are ripple field warnings.

Fear is not weakness. It is the most ancient ripple whisper of consequence.

4. Habits = Rigid Ripples

- Habits are looped ripple behaviors — initially formed for stability, but often become resistance when field conditions change.
- Good habits = adaptive alignment
- Stubborn habits = locked ripple scripts
- Copying others' habits = applying alien ripple patterns to an unmatched field

This reframes habit-building from imitation to personalized ripple learning:

- Not "do what successful people do,"
- But "feel what aligns in your context."

The mind is not a programmable machine. It's a ripple-sensitive instrument — train it to feel, not just repeat.

5. Universal Skill = Real-Time Ripple Perception

- The Universal Skill is not a habit, not a trait, and not talent. It is the evolved capacity to:
    - Read ripple alignment
    - Detect consequences before they arrive
    - Act in harmony with the field's evolving structure

People with high Universal Skill:

- Adapt without panic
- Communicate precisely
- Succeed without manipulation
- Lead without force

This skill cannot be taught traditionally — it must be grown through ripple immersion, failure, silence, and aligned reflection.

The wisest minds don't just think clearly — they feel consequences in advance.

Books Aligned with This Framework

Self-Help: New Syllabus

Reframes growth from motivational habits to consequence literacy — showing readers how to build inner alignment, not discipline.

Out of Syllabus

Challenges conventional learning by revealing the non-linear nature of intelligence, memory, and perception as ripple consequences — not curriculum outcomes.

Live Fearless with Fear

Breaks down fear into ripple behavior, explaining how fear is not to be overcome but realigned with survival intent, restoring clarity and power.

Ripple Mastery

A practical guide for ripple regulation, emotional hygiene, and inner-field maintenance — structured around 12 ripple principles for daily life.

The Human Game

Explores human behavior as a multi-player ripple field, explaining conflict, identity, ego, and growth as field tensions — not personality traits.

Field Applications and Implications

Conventional Psychology	Ripple-Based View
Behavior	Ripple consequence
Emotion	Field turbulence
Motivation	Ripple-pull from misalignment
Trauma	Residual ripple break in memory field
Healing	Realignment, not suppression
Intelligence	Ripple responsiveness and attunement
Mental illness	Long-term ripple distortion or fragmentation

Therapeutic & Development Implications

- Healing becomes re-alignment rather than coping.
- Parenting becomes field modeling, not instruction.
- Leadership becomes ripple coherence, not command.
- Education becomes alignment empowerment, not information transfer.

Closing Thought

You are not your past, your thoughts, or your patterns. You are the ripple field trying to rebalance through them. Observe gently, act attentively, and let your field teach you what the books never could.

Ripple behavior is not a metaphor. It's your real operating system. And once you learn to feel it, everything in life — from anger to love, from failure to insight — begins to make sense again.

Further Reading

Selected works by the author for extended inquiry across ripple science, self-awareness, and universal consequence.

Physics & Cosmology

Books that challenge conventional laws and propose ripple-aligned alternatives through Universal Energy Dynamics (UED):

- **The Law of Truth** → Deconstructs scientific illusion and reframes truth as ripple survivability and consequence persistence.
- **Vishwarup Darshan:** The Complete Vision of Reality → A reimagined model of the cosmos where structured vacuum, ripple balance, and intelligent consequence replace Big Bang cosmology.
- **Manifestation:** From Nothing to Everything → Explores the ripple-based journey of existence — from stillness to awareness, from void to vibration.
- **The Ripple Code (Research Report Series)** → Technical reports on Structured Vacuum, ATT, CTR, and experimental implications of ripple mechanics.

Mind, Emotion & Human Behavior

Works that apply ripple thinking to personal development, psychology, and emotional clarity:

- **Self-Help: New Syllabus** → A radical rejection of habit-driven motivation — offers alignment-based growth rooted in consequence awareness.
- **Out of Syllabus** → Reveals the flaws in formal education and how to learn through field perception, not forced memory.
- **Live Fearless with Fear** → Reinterprets fear as a signal of survival misalignment — not weakness, but wisdom.
- **Ripple Mastery** → A compact handbook of ripple principles for daily awareness, decision-making, and emotional stability.
- **The Human Game** → A field-based decoding of how humans behave in groups, compete, struggle, love, and evolve.

Philosophy, Metaphysics & Consciousness

Books that explore foundational questions using ripple consequence rather than mystical abstraction:

- **The Original Intent** → A reconstruction of ancient philosophies as consequence-aware guidance systems, not ideologies or dogma.
- **The Karma Paradox** → A debate-driven exploration of karma as a ripple law of universal justice — not a

spiritual ledger.

- **The Untold Story of Intelligence and Mind** → Defines intelligence as ripple navigation, not data processing — repositions consciousness as alignment behavior.

Society, Truth & Political Philosophy

Ripple-aware commentaries on systems, identity, power, and moral collapse:

- **Freedom From Politics** → Investigates how ripple misalignment corrupts collective intent — and how to reimagine leadership through clarity.
- **Saffronized:** The Code Behind the Crown → A ripple-based decoding of political transformation in India — from choice to control, from citizen to field-controlled follower.
- **The War of Thoughts (Series)** → A bold series featuring imagined dialogues with great thinkers and scientists, unpacking ripple philosophy across history, science, and spirituality.

Spiritual & Existential Alignment

Books that bridge UED with timeless inner inquiry:

- **Never Speak Louder Than Silence** → A guide to understanding silence not as absence, but as ripple readiness — the space where truth begins.
- **The Equation of Life:** Decoding Dharma for a

Harmonious Life → Translates Dharma into field logic — revealing ethics as ripple-consequence calibration.

Reports & Papers *(Available on Zenodo and select platforms)*

- *Chavan's Cosmic Theory of Relativity (CTR)*
- *ATT: Alignment Threshold Time and Ripple Delay Mapping*
- *Structured Vacuum and the Ripple Reformation of Space*
- *The Soul Demystified: A Field-Based Explanation of Consciousness*
- *Black Hole Ripple Dynamics: Convergence Without Collapse*

These are not books to finish. They are fields to enter. Read them not to agree — but to feel your next ripple realignment.

Acknowledgements

This book was written in silence — but it was never created in isolation.

To my **family**, who may not have read every word but felt every shift: Your patience, your quiet tolerance, and your steady presence allowed this work to unfold without noise. You gave me the space to listen inwardly, and that gift cannot be measured. You believed in my stillness even when you didn't always understand what it was for.

To the **students and thinkers** who unknowingly became part of this inquiry: Your questions, frustrations, and curiosity became subtle mirrors — helping me test my clarity against reality. You kept me honest. You reminded me that truth must be teachable in silence, not just spoken in confidence.

To the **friends who stayed curious** — even if confused — and never asked me to be more conventional than I could bear: Your respect for my madness was a rare kind of encouragement. You didn't need to validate my words — you validated my freedom to find them.

To the **ancient observers and modern minds** whose work I questioned, dismantled, and still deeply respected: Einstein, Newton, Feynman, Krishna, Buddha — and so many others — This book is not a challenge to you, but a continuation of the ripple you began. You taught me to dare. And for that, I'm forever in your debt.

To those who read this and feel something stir — You are the reason this ripple was recorded. You are not followers. You are the next field in which truth may reappear. Thank you for being here, quietly, fully, and freely.

And finally — To the **Field** itself. The structured vacuum. The original silence. The consequence-born intelligence that whispered everything I ever wrote.

I am not its owner. Only its witness.

Thank you.

— *Sandeep*

About the Author

Sandeep Chavan is an independent thinker, writer, and ripple observer whose work spans physics, cosmology, psychology, philosophy, and human behavior — all unified under the field-aware lens of Universal Energy Dynamics (UED).

With no formal lab, no institutional label, and no interest in academic conformity, Sandeep has spent thousands of solitary hours decoding the universe from the inside out — not as a scientist chasing credentials, but as a seeker of structural truth. His insights are consequence-based, alignment-driven, and unapologetically original.

He has authored over 100 books and research papers, exploring everything from ripple mechanics and time perception to emotional intelligence, karma, silence, and the soul. His works reject conventional silos, blending rigorous observation with inner clarity to create a new kind of science — one that breathes.

A teacher by profession and a philosopher by consequence, Sandeep lives and works in India, where he continues to write, teach, and observe — far from institutions, but deeply aligned with reality.

He doesn't want followers. He writes for feelers — the ones who sense first, and believe later.

ORCID Profile: https://orcid.org/0009-0005-4190-7164

Disclaimer

This book is a personal intellectual exploration and a reflective narrative based on the author's independent research, philosophical inquiry, and observational synthesis. It is **not a textbook**, academic publication, or scientific manual validated by institutional peer review.

The concepts presented here — including Universal Energy Dynamics (UED), ripple behavior, alignment theory, and field-based interpretations of physics, cosmology, psychology, and philosophy — are **original interpretations and frameworks developed by the author** and are intended to provoke thought, inquiry, and individual reflection.

Readers are encouraged to engage with these ideas as a **new lens of understanding**, not as absolute replacements for existing scientific models or psychological paradigms.

This work should not be used as a substitute for formal education, professional scientific consensus, medical or psychological advice, or institutional research. The author makes no guarantees about the acceptance, implementation, or outcomes of these concepts in academic, clinical, or applied settings.

All references to historical figures, scientific theories, or philosophical schools are made with **deep respect**, and no infringement or misrepresentation is intended.

Don't miss out!

Visit the website below and you can sign up to receive emails whenever Sandeep Chavan publishes a new book. There's no charge and no obligation.

https://books2read.com/r/B-A-EPGPC-ZFGJG

BOOKS 2 READ

Connecting independent readers to independent writers.

Did you love *Feeling Like a Physicist*? Then you should read *Cracks in the Known*[1] by Sandeep Chavan!

[2]

Cracks in the Known

Redefining Reality through Science, Logic, and the Nature of Truth by **Sandeep Chavan**

What if the truths we've accepted for centuries are only shadows on the wall?What if time doesn't pass, gravity doesn't pull, and the universe never really began?

In *Cracks in the Known*, independent thinker and philosophical disruptor **Sandeep** invites readers into a dreamlike intellectual arena—where the greatest minds in

1. https://books2read.com/u/b6XOlA

2. https://books2read.com/u/b6XOlA

history return, not to lecture, but to defend their legacy against a new wave of inquiry. Through imagined debates with legends like Einstein, Plato, Newton, Penrose, and Krishna, Sandeep dismantles the foundations of science and philosophy—not out of rebellion, but to reveal something deeper beneath them.

This is not a book of answers.It is a confrontation with the assumptions behind the answers.

From the illusion of time to the fiction of force, from the paradox of origin to the layered nature of energy itself, *Cracks in the Known* doesn't just challenge accepted frameworks—it replaces them with new ones, rooted in balance, awareness, and the hidden intelligence within the universe.

At the center of the narrative is Sandeep's own theory: **Universal Energy Dynamics (UED)**, a model that recasts gravity, time, and even creation as consequences—not causes. Here, energy is not inert; it is conscious. And what we call "truth" may be nothing more than equilibrium expressing itself through experience.

This is not conventional science.It is not abstract philosophy.It is a **new form of thinking**—one that merges logic with silence, and intellect with intuition.

Every chapter is a war.Every conclusion is a door.And by the final page, readers may find themselves not at the end of a book—but at the edge of what they once thought was real.

Sandeep doesn't aim to convince.He aims to awaken.

Because the cracks in the known... were always there.We just stopped looking.

Read more at https://www.sandeepjchavan.com/.

Also by Sandeep Chavan

Chavan's UED research

CMGED Research Report

Chavan's Cosmic Continuum (C^3)

Frequency

Ripples of Reality

Reality Realigned

The Great Realignment

I – The Cosmic Leverage

Chavan's Cosmic Theory of Relativity (CTR)

Chavan's Theory of Light

Feeling Like a Physicist

Echoes of War: The Russia-Ukraine Conflict and Its Global Ripples

Shattered Lives

The War Economy

Borders Redrawn

Arms and Allies

Collateral Damage

Justice in the Ashes

Somewhere in the Darkness of the Cosmos

The Blueprint of the Universe

The Cosmic Breath

The Eternal Sage Chronicles

The Sage and the Sovereign

The Family Pillar

The People's Flame

The Warrior's Soul

The Bridging Mentor

The Guiding Light

The Exodus

The Last Cycle: 2075

The Fundamental Theory of Life & Existence

Energy in Action: The Soul Demystified

The Active Mind

Energy in Connection

Death: The Disconnection

Subconsciousness: The Chosen Disconnection

Desires: The Engine of Life

The Restoration of Human Knowledge

Restoring Bhagavad Gita

The War of Thoughts

Cracks in the Known

The Human Game

The Divine Dilemma

The Original Intent

The Answers

Ultra Modern Life Philosophy

The Duality of Laughter and Tears

Take It Easy

Compare to Accept

Standalone

It's Not AI, It's AHI - Amplified Human Intelligence

The Decision Paradox

The Four Sapiens

Malicious Script of Indian Polity

The IIT Legacy & Global Impact

The IIT Legacy & Global Impact

You are just a Version of Your Original

Win the Game You Didn't Choose

Always Keep Your Bags Packed

Reveal to Shield - Learn the Game of Facades

Engage Beyond Elections

Culture, Identity & Change: The Evolution of Indian Society

The Triangular Dynamics

Never Trust Without Doubt

If You Can't Change the Country, Leave the Country

Live Fearless with Fear

India: The Investment Magnet

Realignment

Powering a Billion Dreams

Universal Skill

Extension: From Brush to Algorithm

Human Side of Einstein

The Future Knows No Lines

Let's Bring the Generations to the Table

Unified Energy Dynamics

A Beat That Never Dies

Strategic Horizons

Right From Scratch

The Forgotten Innovators

Physics Fixing

The Savior Illusion

The Selfish Side

The Age of the Weakest Sapiens

Neural Wars

The Absorption: There Is No Death

Strange Therapist

Bubbles: Become a Drop

Beyond the Blue

Half-Scholar

The Law of Truth
Failure
Philosophy 2.0: A Structural Manifesto
Chaos: A Gentle Guide to the Unresolvable
Self-Help: New Syllabus
Rippling
Game: How to Play the Share Market
Out of Syllabus
Saffronized: The Code Behind the Crown
The Anger

Watch for more at https://www.sandeepjchavan.com/.

www.ingramcontent.com/pod-product-compliance
Lightning Source LLC
LaVergne TN
LVHW090406210926
848480LV00002B/466